Advance Praise for *Serious Microhydro*

Microhydro is the renewable of choice for anyone with available flow, or if you're just curious about the culture. Scott Davis' book summarizes much of the cumulated wisdom. It's required reading for anyone thinking about going hydro. Practical; clearly written. I was fascinated.

Paul Craig
Former Chair,
Sierra Club National Global Warming
and Energy Committee

We at PowerPal are delighted to see Scott's new book, *Serious Microhydro*, as we are sure that it will quickly become the "go to" reference book for anyone seriously contemplating a project to generate electricity from the energy of running water. It is an excellent follow-on from his first book, and the many case studies are both factual and fascinating. Again, we will be recommending this new book to many of our clients.

David Seymour
President & CEO of
Asian Phoenix Resources Ltd.

Scott Davis brings together an excellent collection of microhydro experiences. *Serious Microhydro* provides an essential source of information for those interested in developing microhydropower, and for anyone interested in learning more about it. This collection clearly shows the importance of hydropower in providing renewable energy for a sustainable future.

Ghanashyam Ranjitkar
Hydraulic Energy Engineer, Marine Energy Group
Sustainable Buildings and Communities-RET
CanmetENERGY
Natural Resources Canada

serious
microhydro

WATER POWER SOLUTIONS FROM THE EXPERTS

edited by

Scott L. Davis

NEW SOCIETY PUBLISHERS

Cover design by Diane McIntosh.
Water splash: iStock/plainview; Insets (from left to right): Kurt Johnson,
Peter Talbot, Jim Lakey, Roy Davis; Spine: Bob Mathews

Printed in Canada.
First printing August 2010.

New Society Publishers acknowledges the support of the Government of Canada
through the Book Publishing Industry Development Program (BPIDP)
for our publishing activities.

Inquiries regarding requests to reprint all or part of *Serious Microhydro*
should be addressed to New Society Publishers at the address below.

To order directly from the publishers, please call toll-free
(North America) 1-800-567-6772, or order online at
www.newsociety.com

Any other inquiries can be directed by mail to:

New Society Publishers
P.O. Box 189, Gabriola Island, BC V0R 1X0, Canada
(250) 247-9737

New Society Publishers' mission is to publish books that contribute in fundamental
ways to building an ecologically sustainable and just society, and to do so with the
least possible impact on the environment, in a manner that models this vision. We are
committed to doing this not just through education, but through action. This book is
one step toward ending global deforestation and climate change. It is printed on Forest
Stewardship Council-certified acid-free paper that is **100% post-consumer recycled**
(100% old growth forest-free), processed chlorine free, and printed with vegetable-
based, low-VOC inks, with covers produced using FSC-certified stock. New Society
also works to reduce its carbon footprint, and purchases carbon offsets based on an
annual audit to ensure a carbon neutral footprint. For further information, or to browse
our full list of books and purchase securely, visit our website at: www.newsociety.com

Library and Archives Canada Cataloguing in Publication

Serious microhydro : water power solutions from the experts / edited by Scott Davis.

Includes index.
ISBN 978-0-86571-638-4   eISBN 978-1-55092-448-0

1. Water-power.   2. Small power production facilities.
3. Hydroelectric power plants.   I. Davis, Scott (Scott L.)

TK1081.S47 2010   621.31'2134   C2010-904021-X

NEW SOCIETY PUBLISHERS
www.newsociety.com

Mixed Sources
Cert no. SW-COC-001271
© 1996 FSC

FSC

DEDICATION

I'd like to dedicate this book to my family,
who make it all possible:

To my wife and partner of many years, Bonnie Mae Newsmall
(with her hand on the valve in the photo),
and our daughter Alannah
(standing just behind me in the hat).

And of course to my parents:
Roy, the photographer here whose expertise at things mechanical
saved this initial start-up of this, my first microhydro system.
And to my mother Della (behind me there, as ever),
who died June 30, 2008.

Many hands make lights work!

Roy Davis

Contents

Part V: The Future — Bringing Market Solutions to Environmental Problems

Acknowledgments

So many people made this possible. I'd like to thank New Society Publishers and those there who have been so patient with me while turning out such a beautiful product:

Chris Plant, Judy Plant, Ingrid Witvoet, Betsy Nuse and all the rest of the crew, my thanks.

I'd also like to thank Fred Brown for offering the opportunity, among so many other things, of working on a microhydro system, all those many years ago. There's nothing like developing a project to teach you what you need to know about energy issues

And there is nowhere better to go for firsthand accounts of this process than *Home Power* magazine. My thanks for their inspiration over the years and for their help with this book. And of course, I would like to thank the authors of these chapters, both for their life experience and for the time they took to tell the tale: Jeffe Aronson, Barbara Atkinson, John Bethea, Paul Craig, Paul Cunningham, Ron Davis, the ESMAP Project of the World Bank, Jim Forgette, Roger and Linda Gaydos, (the late) Stephen M. Gima, Eileen Loschky nee Puttre, Chris Greacen, Jeevan Goff, Jo Radtke nee Hamilton, Arne Jacobsen, Juliette and Lucien Gunderman, Philippe Habib, Don Harris, John Hermans, Paul Hoover, Kathleen Jarschke-Schultz, Kurt Johnson, Chester Kalinoski, Bill Kelsey, Terry Kinzel and Sue Ellen Kingsley, Dan Loweburg, Cameron MacLeod, Bob Mathews, Jerry Ostermeier, Dave Palumbo, Richard Perez, Harry O. Rakfeldt, Auden Schendler, Bob-O Schultz, Joe Schwartz, David Seymour, Hugh Spencer, Phillip Squire, Peter Talbot, Malcolm Terrence, Ian Woofenden, Louis Woofenden and Rose Woofenden.

Scott L. Davis
March 2010

Introduction

Using This Book

Serious Microhydro brings together dozens of perspectives on the experience and context of generating power from small water-powered alternators. While microhydro can be the most cost-effective renewable energy technology, it is still challenging to find relevant material to learn about the topic. For whatever reasons, microhydro has not been marketed nearly as well as photovoltaics or wind power. There has never been a market study of microhydro. So, any assumptions or generalizations may only reflect individual experience or prejudice. This collection represents dozens of viewpoints over decades and could be considered the next best thing to such a market study. *Serious Microhydro* will help us understand important elements of current microhydro practice — such as the ratio of systems with or without batteries — or even what an average or typical system might look like.

Testimonials are powerful teaching and sales tools. Microhydro is like other renewable energy technologies in that its proponents must convince prospective customers that a high standard of living can be possible with much less power than they are accustomed to using. Here the reader can find dozens of specific examples on this and a number of other important questions about the context of microhydro in renewable energy:

- How is water flow estimated?
- What are the costs of a project?
- How much power is enough for a high standard of living?
- What part of an energy budget could be served with heat, and what part of the budget actually needs electricity?
- How much drop is necessary to generate power?
- What renewable energy incentives work best?
- How is debris kept out of a turbine?
- How can an airlock be cleared from a pipe?
- What is the best way to share power between households?
- Can surplus power be sold?
- How far can the turbine be from the point of use?
- Where can money be saved — or better — where and how should a limited budget be spent?
- How can occasional surpluses of power be used?
- What are the barriers to microhydro project development where utility power is available?
- What does the future hold in store for microhydro?
- How can people get the experience with

renewable energy issues in general, and with microhydro issues in particular, that is needed to make informed decisions?

Stand-alone power is usually more expensive per kilowatt than utility power. In small projects economies of scale are lost, but the pricing to provide a stand-alone power supply also approaches ideal full-price accounting much more closely than utility rates do. The microhydro project developer isn't externalizing environmental and social costs. That developer receives none of the outright subsidies which

---

...if regular affluent North Americans realized that a high standard of living only really needed a quarter of their current power consumption, that would be the end of the energy crisis as we know it, would it not?...

---

distort the costs of on-grid projects. Off-grid, the user pays for capacity, timing and quality of power in a way that makes energy efficiency sensible. The point is that a high standard of living is possible with a fraction of current consumption. People who build their own power systems learn valuable lessons about energy realities which the reader can share. These people find out that clean energy for a high standard of living is affordable. If everyone realized the full implications of this basic truth, there would be little reason for an energy crisis.

So, precisely as you have heard people say for decades, you just can't afford to waste. What *is* affordable from clean power is a high standard of living. What is not affordable is the kind of waste that happens when energy prices

are subsidized by externalizing environmental and social costs.

To solve the energy crisis and get a stand-alone power system that is big enough without being too big, affluent North Americans must realize that a high standard of living requires but a quarter of their current power consumption. This means that we will recognize that common appliances are incredibly inefficient, and that we've continued to use them only because we have never paid the full price for power (or many other commodities, for that matter). Now that compact fluorescents are more common, demonstrating this point is easier.

Confusion, however, reigns as a result of distortions brought about by low energy prices and poor environmental standards. For example, as renters we have a refrigerator that proudly claims to use 145 kilowatt-hours per month. By contrast, a Danish refrigerator (the Vestfrost) uses about 30 kilowatt-hours per month. Danes and many others pay a lot for residential power. There are many examples, from lighting to refrigeration, in which North Americans use many times as much energy as necessary. Even today, Canadian hospitals use six times as much power as Swiss hospitals even though Swiss electricity is only about twice as expensive.

Power from photovoltaic panels is relatively expensive in terms of upfront costs. So it is straightforward to demonstrate that it would be a much better idea to buy a US$4,000 refrigerator that only uses 16 kilowatt-hours a month than to buy a dozen or more panels at US$1,000 each and perhaps upgrade the battery/inverter subsystem to generate the additional power necessary to start and run a less efficient refrigerator. Many of the articles in this book have detailed load analyses that show how, with efficient equipment, the wide variety of modern conveniences can be provided from an astonishingly small electrical power output.

Microhydro, where available, is the most cost-effective renewable energy technology. Thus, microhydro systems are relatively more powerful from the standpoint of kilowatt-hours per dollar invested. In practice, this means that even an average-sized microhydro system can put out a remarkable amount of power, surpassing all but the largest photovoltaic systems. So anything a PV-powered system can do, like charge up electric vehicles, can be done with a much less expensive microhydro system — if, of course, the water power resource is available.

In fact, most of the microhydro systems described in this book produce enough power that provisions for dumping the excess are required. My own system, described here in "The Small AC System" (Chapter 5), had baseboard heaters on the back porch to dump excess power in the summer. More power could have been generated from the site, but we found that without a grid to sell to, the amount that a single household could use, even as free heating, had its limits.

Always keep both costs and benefits in mind. In stand-alone situations, the benefits from producing more power show distinctly diminishing returns. At the same time, costs in these situations go up pretty much in a linear fashion, with a larger system costing somewhat more than a smaller one.

How Much Power is Enough?

There is a rule of thumb amongst microhydro developers, repeated more than once in these articles, that a system producing 300 continuous watts, and with the appropriate battery/ inverter subsystem, will provide modern conveniences at reasonable cost for the off-grid household. In this book you'll also find many examples of people doing very well on even less. Even better efficiency is possible and affordable. If you listen to the users of photovoltaic and wind power systems, they say that the first 100 kilowatt-hours per month is essential for a high standard of living.

These essential kilowatt-hours are costly to provide with other technologies. Although it is commonly said that diesel-powered generators can produce power for 40 cents a kilowatt-hour, small stand-alone units, powering household loads and maintained by their owner, seldom get this cost-efficient. Using a generator for light loads such as illumination increases maintenance cost. Diesels need to be quite heavily loaded to be efficient, and lights just don't provide this kind of load.

As an off-grid project developer, I found many a heap of dead generators at otherwise green and organic homesteads, each generator bigger and more expensive than the last. Costs can easily double or more when maintenance and depreciation are figured in. That's why people like PV systems, even when power from such a system costs a dollar or so a kilowatt-hour.

Since PV systems work more or less everywhere to provide power, figuring that the first 100–200 kilowatt-hours per month might cost up to a dollar apiece is not unreasonable. Since there are approximately 720 hours in a month, then a microhydro system that produces as little as 140 continuous watts will provide 100 kilowatt-hours in a month.

A system producing 200 kilowatt-hours per month is quite handy, if only because more ordinary appliances may be used. Unlike photovoltaics, the costs of producing a larger microhydro system may not be as directly related to the output. Sometimes, more power can be produced with relatively little expense, like going uphill a little farther, using a little more water with bigger pipe or by getting more efficient technology.

Then When Do We Quit?

Off-grid, power outputs over the level of a couple of hundred kilowatt-hours per month are used to provide heat, one way or the other. Inefficient appliances heat the air, but this energy can also be captured to heat water or in space heating. This kind of power is less valuable, since low grade energy demands can be met with a variety of technologies that are less expensive. These include insulation and solar water heating. Consider kilowatt-hours in excess of 300 to be much less valuable to the stand-alone system, certainly half or less. In addition, demand is limited. There is only so much hot water that even the most enthusiastic household can use, say another couple of hundred kilowatt-hours a month, and then power is used for space heating.

Space heating can consume thousands of kilowatt-hours per month in a cold climate, but again, this energy is even less valuable. Many heating *requirements* are more the consequences of poor design and lack of insulation. The value of kilowatt-hours in excess of 1,000 or so per month is a quarter or less of the value of the first couple of hundred kilowatt-hours.

There is another innovative and important use for the output from larger systems, which is to power electric vehicles. Although the deep backwoods off-grid kind of site where microhydro systems are commonly found is often not the best place for electric vehicles, the fact that these vehicles use about 300 watt-hours per mile has to be one of the best uses for the output of larger systems. The possibility of free fuel indefinitely from your water system has got to be a significant incentive.

Why take my word for these assumptions? Here, in *Serious Microhydro*, you can go directly to primary source after primary source to find testimonials about how well systems meet peoples' needs. Some of the very small systems seem to meet needs very well.

What Does the Typical Microhydropower System Look Like?

Compare and contrast the specifications of the dozens of sites in Figure 1: Case Studies by Head. Note that the median size system here is about 200 watts, although this is relatively distorted by the large number of ESMAP (Energy Sector Management Assistance Programme, Ecuador) case studies which used systems that size. It does say something about minimum system size and performance, however. If you only look at North American sites, the average output is about 400 watts. This is less than half of the average consumption rate on-grid, yet nowhere will you get the idea that people are sacrificing very much.

This chart can help you visualize a potential site. To understand the issues involved, just find a case study here that has a similar *head* (the drop that water undergoes to produce power). Imagine you have a nearby creek to develop with 50 feet of head available. Look at sites in the figure with about 50 feet of head. Since size is money, seeing how others meet their needs can be most enlightening. Remember, too, that there's no rule that you have to use all the head available. With an adequate flow of water, there may be much more convenient solutions using five or ten feet of head. So then, a place with a total of 50 feet of head might have numerous possible five or ten foot sites.

The heart of this book is in the dozens of case studies which illustrate the range of solutions to providing power from local resources. This range is astonishingly wide because sites vary so much in head, flow and output — plus every site has many other unique elements. In *Serious Microhydro*, heads range from 1.9 feet to 746 feet. The flows of water from one site to another vary from under ten US gallons per minute to over 17,000. Outputs range from 30 kilowatts, powering a wilderness lodge, down to 18 watts, providing a light or so for a remote household. Different people make different demands upon the resource. So, at any given site there may be many different kinds of technology that might be appropriate.

The case studies are divided into three groups, according to the static heads available at the site, and thus, also grouped according to the technology involved. The classic microhydro site — the one most people would think about when they hear the word — would be a powerful, high head AC site in rainy mountains. These classic sites comprise the group in Part I.

However, similar if smaller technology can also extract power from much smaller gravity water systems, such as provide water to remote households. These systems might in addition use a battery/inverter subsystem, like a PV or wind system does, to make much from little. This remarkably large and various bunch comprises Part II.

Part III groups together low head sites. Generating power from the volume rather than from the pressure of the water is done with a variety of technologies, from the overshot waterwheel to the propeller-type turbine. Pay close attention to the possibilities of innovative low head technology.

Part IV addresses various issues involved in developing a successful system — from a classic introduction to the topic, to history, to technical hints, troubleshooting techniques and innovative solutions to some design problems. What will not be found here are any of several good, detailed, introductory articles about microhydro in the literature. Just as there is little agreement about simple things like nomenclature, there are surprising and confusing contradictions from one introduction to another.[1] It's much better to hear people discuss their own experiences, as they do here, and take what you need from it. The classic "Microhydro Power in the Nineties" (Chapter 35) introduces elements that are essential to designing and operating a successful system.

Part V concerns the future of microhydro. The rise of energy farming, net metering and grid intertie means that there are wonderful future opportunities for microhydro where utility power is available. All that seems to be needed are appropriate incentives, ones that make renewable energy practical.

At this time, microhydro still performs best in the off-grid environment. In a world where so many people are without electricity and are spending a lot of the dollar or two a day they earn on kerosene, appropriate incentives can bring power to those with none. Part V concludes with the proposal for the Stream-works project, which seeks to provide microfinancing for microhydro.

However you use this book, good luck! And now, on to the projects themselves.

Case Studies by Head

Static Head in Feet	Chapter	Location	Flow (in US gallons per minute)	Output (in watts)	Type	Use
1.9	24 System 2	US	3,890	390	130 volt DC Battery	Single family full-time residence
3	24 System 3	US	17,640	4,550	130 volt DC Battery	Single family full-time residence
4	24 System 5	US	16,382	7,000	250 volt DC Battery	Single family full-time residence
4	29	Australia	3,200	1,200	240 volt AC Battery/inverter	Uses pumped storage and multiple turbines
5	24 System 1	US	539	390	130 volt DC Battery	Single family full-time residence
5	30 Units #6120 and 6353	Ecuador	1,100 (2×550)	400 (total)	220 volt AC	Note use of two turbines to service two house-holds
5	30 Unit #6121	Ecuador	550	200	220 volt AC	Serves three households with four 20 watt lamps and two radios in total
5	30 Unit #6123	Ecuador	950	500	220 volt AC	Powers a demonstration site
5	30 Unit #6125	Ecuador	550	200	220 volt AC	Provides power for a house
5	30 Unit #6127	Ecuador	550	200	220 volt AC	Powers four households each with a 20 watt lamp and a radio
5	30 Unit #6128	Ecuador	550	200	220 volt AC	Powers four households each with a 20 watt lamp and a radio

Scott L. Davis

Static Head in Feet	Chapter	Location	Flow (in US gallons per minute)	Output (in watts)	Type	Use
5	30 Unit #6131	Ecuador	550	200	220 volt AC	Powers a household with two 20 watt lamps and a radio
5	30 Unit #6134	Ecuador	550	200	220 volt AC	Powers four households each with a 20 watt lamp and a radio
5	30 Units #6136, 6355, 6130 and 6132	Ecuador	2,200 (4×550)	800 (total)	220 volt AC	Note use of four turbines to service four households
5	30 Unit #6138	Ecuador	550	200	220 volt AC	Powers a household with three 20 watt lamps and a radio
5	30 Unit #6350	Ecuador	550	200	220 volt AC	Powers four households each with a 20 watt lamp and a radio
5	30 Unit #6351	Ecuador	550	200	220 volt AC	Powers three households and a poultry breeding hut
5	30 Unit #6352	Ecuador	550	200	220 volt AC	Provides electricity for a household
5	30 Units #6356, 6341 and 6122	Ecuador	1,660 (3×550)	600 (total)	220 volt AC	Note use of three turbines to service three households
5	30 Unit #6357	Ecuador	550	200	220 volt AC	Powers a school, teacher's house, workshop and community center
5	30 Unit #6359	Ecuador	550	200	220 volt AC	Powers three households (one 20 watt lamp and a radio each) and a cheese factory (one 20 watt lamp and a radio)

Static Head in Feet	Chapter	Location	Flow (in US gallons per minute)	Output (in watts)	Type	Use
6.5	27	Australia	1,800–3,000	250–300	24 volt AC Battery/inverter	Single family full-time residence
6.5	28	Australia	800	450	120 volt DC Battery/inverter	Single family full-time residence
10	24 System 6	US	1,825	1,950	130 volt DC Battery	Single family full-time residence
14	26	Northern California	449	2,500	120 volt AC	Single family residence and farm
15	24 System 4	US	972	1,560	130 volt DC Battery	Single family full-time residence
16	23	Michigan	75	115	12 volt DC Battery/inverter	Single family residence
17	25	US	5,400–30,000	12,000–30,000	120/240 volt AC	Two full-time residences and shop
20	30 Unit #6550	Ecuador	95	200	220 volt AC	Demonstration unit that powers two 20 watt lamps and a radio for a small household inside the training institution
20	30 Unit #6551	Ecuador	95	200	220 volt AC	Powers two households; one is an ecolodge
20	30 Unit #6557	Ecuador	95	200	220 volt AC	Powers the main hall and kitchen of a church plus two households each with a 20 watt lamp and a radio
26	30 Unit #6555	Ecuador	95	200	220 volt AC	Powers a household with two 20 watt lamps and a radio
26	30 Unit #6559	Ecuador	95	200	220 volt AC	Powers a household with two 20 watt lamps and a radio

Static Head in Feet	Chapter	Location	Flow (in US gallons per minute)	Output (in watts)	Type	Use
27	22	Northern California	35	50	24 volt DC Battery/inverter	Single family residence and business
30	20	Connecticut	80	150	24 volt DC Battery/ inverter	Single family residence
30	19	Washington	160	525	48 volt DC Battery/inverter	Single family residence
35	18	Eastern US	20	21	12 volt DC Battery	Weekend cabin
40	30 Unit #6553	Ecuador	95	400	220 volt AC	Residence
40	17	Northern California	12	18	12 volt DC Battery/inverter	Single family full-time residence
40	16	Northern California	9.7	18	12 volt DC Battery/inverter	Eichenhofer full-time residence
55	*Home Power* #71 (June/ July 1999)	Costa Rica	73	410	28 volt DC Battery/inverter	Induction generator for Lebofamily full-time residence
55	*Home Power* #71 (June/ July 1999)	Costa Rica	300	380	258 volt AC Battery/inverter	Induction generator for Llano full-time residence
65	12	Bolivia	82	"similar to ¾ horse-power motor"	Mechanical drive	The Watermotor powers tools directly from the turbine
65	*Home Power* #71 (June/ July 1999)	Costa Rica	185	630	334 volt AC Battery/inverter	Induction generator for Stewart/Bomene residence
72	16	Northern California	10	33	12 volt DC Battery/inverter	Hanauer family full-time residence

Static Head in Feet	Chapter	Location	Flow (in US gallons per minute)	Output (in watts)	Type	Use
75	15	North Carolina	65	400	57 volt DC Grid intertie inverter	Grid intertie
78	14	Nicaragua	21	125	12 volt DC Battery	Central battery shed provides lights and other small loads for a village of nine households
81	*Home Power* #71 (June/ July 1999)	Costa Rica	55	375	415 volt AC Battery/inverter	Induction generator for Buena Vista lodge
100 (est.)	6	Northern California	20	120	12 volt DC Battery/inverter	Pullen family full-time residence
100	11	Queensland Australia	9.5	30	12 volt DC Battery/inverter	Residence and research station
103.5	10	Ashland Oregon	45	288	24 volt DC Battery/inverter	Household and business
112	5	Canada	550	3,200	120 volt AC	Provides considerable space heating for ranch as well as many modern conveniences
138	6	California	90	2,500	120 volt AC	Gene Strouss full-time residence
148	*Home Power* #71 (June/ July 1999)	Costa Rica	60	410	415 volt AC Battery/inverter	Induction generator for home and lodge under construction
161	8	US	22	430	24 volt DC Battery/inverter	Powers a home and turbine business
170/ 85	4	McMinnville Oregon	500	4,100	240 volt AC Grid intertie	Independent power producer (two turbines, one alternator)
173	*Home Power* #49, (Oct/ Nov 1995)	New Zealand	N/A	68	17 volt DC Battery/inverter	Hybrid with PV system

Static Head in Feet	Chapter	Location	Flow (in US gallons per minute)	Output (in watts)	Type	Use
175	6	California	10	97	12 volt DC Battery	Creasy family full-time residence
180	6	California	15	180	24 volt DC Battery/inverter	Stan Strouss full-time residence
180	13	California	9	150	24 volt DC Battery/inverter	Retreat Centre
190	*Home Power* #71 (June/ July 1999)	Costa Rica	21	180	415 volt AC Battery/inverter	Induction generator for Bosque Del Cabo lodge
200	9	Oregon	35	600	24 volt DC Battery/inverter	BLM guest ranch demonstration site 24 volt upgrade
200	9	Oregon`	45	168	24 volt DC Battery/inverter	BLM guest ranch 12 volt system
210	7	Vermont	38	430	200+ volt AC Battery/inverter	Single family full-time residence
280	6	California`	110	3,000	120 volt AC	Gary Strouss winter system
280	6	California	10	120	12 volt DC Battery/inverter	Gary Strouss summer system
315	3	Golden BC	220	12,000	240/600 volt AC	Lodge
328	29	Australia	30	1,200	240 volt AC	Pumped storage for low flow
328	29	Australia	8	120	12 volt DC	Very low flow option pumped storage
550	2	BC	225	12,600	240 volt AC	Club
746	1	Colorado	1,100	30,000	480 volt AC Grid intertie	Independent Power Producer/net metering

PART I

Classic High Pressure Sites

Classic microhydro sites look very much the way people expect water power technology to look. Towering mountains and rainy climates make microhydro a natural source of power. Here, gravity water systems are common and may be put to use making hydroelectricity for sale to the local grid, as they do in Chapters 1 and 4. More commonly, these sites power off-grid lodges and clubs (Chapters 2 and 3).

Sustainable Skiing—
Snowmass Ski Area Gets Hydro

AUDEN SCHENDLER

The histories of Aspen, Colorado, and hydro-electricity converge underground. Silver lodes drew the miners who first established Aspen. And Lester Pelton, the inventor of the modern waterwheel, was a gold miner in California. Both were pursuing a holy grail—vast wealth from the earth's natural resources.

The silver miners found it in Aspen, once in the form of a 2,200 pound silver nugget. Pelton discovered no gold, but he extracted something more valuable—an efficient way to make clean energy from falling water. One hundred and forty years later, his invention, the Pelton wheel, is being put use in a ski resort near Aspen in a revolutionary way.

Sustainable Vision

The silver lodes are long since tapped out, but there is a new grail, of sorts, for the residents of this resort town. It is the idea of a sustainable community, one that can thrive with minimal impact on the environment. In the big picture, the main barrier to that vision is energy use.

As Vijay Vaitheeswaran points out in *Power to the People*, his superb book on global energy issues, "The needlessly filthy and inefficient way we use energy is the single most destructive thing we do to the environment." The average US household is responsible for the annual emission of 23,380 pounds of carbon dioxide, the primary greenhouse gas, much of that from electricity use. Now, consider the emissions from plugging in a ski resort. And yet, "With enough clean energy,"

The microhydroelectric plant on Fanny Hill now has an educational display that will be viewed by an estimated 750,000 skiers annually.

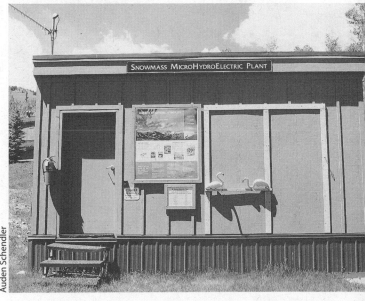

Vaitheeswaran notes, "most environmental problems — not just air pollution or global warming, but also chemical waste and recycling and water scarcity — can be tackled."[1]

The Pelton Wheel

In 1864, when Lester Pelton worked in the mines, mechanical power came from waterwheels spun by jets of water. As the technology evolved, millwrights replaced wooden slats with metal cups, which turned the wheel faster. One day, Pelton observed a broken waterwheel. The jet was hitting the edge of the cup instead of the center. Pelton observed something else — the wheel turned faster than other wheels nearby. Based on his observations, Pelton developed a more efficient design and patented it. That design became the key

Water from the turbine exits the tailrace.

Auden Schendler

component of many modern hydroelectric turbines. A Pelton wheel looks like an industrial flower or a blacksmith's rendition of the universe. It is a beautiful and timeless tool, a reminder of human ingenuity that evokes the creativity of a silversmith more than the equations of an engineer. Pelton wheels have brought great affluence to the world through the sale and use of electricity — and great environmental damage through the construction of large dams. But the first wheel that Lester Pelton put to practical use ran his landlady's sewing machine. Now, that legacy is helping to stitch together the fabric of a sustainable community.

Why Hydro?

Aspen Skiing Company, which operates four ski mountains — Aspen, Snowmass, Highlands and Buttermilk — and several hotels, is responsible for 28,000 tons of greenhouse gas pollution every year. Roughly 23,000 tons of that is from electricity use. One of the only ways to address this impact is to buy renewable electricity, which anyone (even homeowners) can purchase from the local utility, Holy Cross Energy.

The city of Aspen buys 67% of its electricity as renewables. Aspen Skiing Company buys wind power — about 5% of total usage — and increases its purchases annually. But the business can't afford to buy renewables in the volume necessary to offset impacts, and the practice sometimes confuses guests. The most common question is, "Where's the windmill?"

Installing a wind turbine on-site would be a significant investment. The best sites are far from transmission lines, on the local ridgetops. Areas closer to the transmission infrastructure are more sheltered, so there's not enough wind. Photovoltaic panels are an option, but they're expensive, especially for the quantity of energy required. However, one source of renewable energy on ski hills is plentiful, economical and readily at hand — water.

Early Aspen

Early Aspen was all hydro-powered. In fact, according to *The Electric Review* from January 1907, "Aspen led the way in the use of electricity for domestic lighting and mining. For years, it was the best-lighted town in the United States. It was the first mining camp to install an electric hoist, and the first to install generators run by water power."

Today, three substantial microhydro systems are still running in the area (and likely many smaller ones). One is on Maroon Creek and puts out 450 to 500 kilowatts. A 20 kilowatt system is in the basement of the Mountain Chalet in Snowmass. And local microhydro enthusiast Tom Golec has a 40 kilowatt turbine on Ruedi Creek. Unlike dams, microhydro plants take some of the water out of the creek, but don't have to block the flow. Such systems can generate electricity from relatively small streams — you don't need to rebuild the Hoover Dam. The water runs through a pipe to the turbine and then back into the creek downstream.

A Not-So-Costly Installation

The biggest expense of most microhydro systems is the *penstock*, or pipe, that runs from high elevation to low, creating pressurized water that can spin the Pelton wheel. The economics of installing a penstock can often kill a project. At Snowmass Ski Area, installing

Technical Specs

Location: Fanny Hill, Snowmass Ski Area, Snowmass, Colorado
Owner: Aspen Skiing Company
Project cost: US$155,000
Head: 746 feet
Pipeline length: 4,103 feet
Static pressure at turbine: 323 pounds per square inch
Average flow: 1,100 gallons per minute
Turbine: Single nozzle Pelton turbine from Canyon Hydro, 18.5 inch pitch diameter
Generator: 175 horsepower, 480 volt, three phase, 60 hertz, 115 kilowatt
Annual generation: 250,000 kilowatt-hours, estimated

The Pelton wheel used in the Snowmass Ski Area hydro plant was custom-made for the project by Canyon Hydro.

Auden Schendler

Auden Schendler

Snowmass Microhydro System

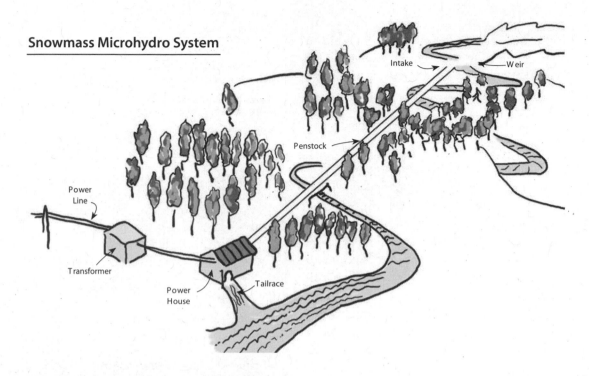

Intake · Weir · Penstock · Power Line · Transformer · Power House · Tailrace

Frank White, doing repairs

Thierry Burkhart

a basic hydroelectric system would require building a retention pond (at a cost of about one million US dollars) and burying 4,000 feet of 10 inch steel pipe. The cost of such a project is mind-boggling. Once you add up pipe cost and excavation equipment time, you're pushing a system's payback into the next millennium. Unless, of course, you have the pipe and pond already in place. At the Snowmass Ski Area in Aspen, we do. We call it a snowmaking system.

Snowmaking pipes run everywhere at some ski resorts. So snowmaking supervisor Jimmy Holton asked, "If we already have half a hydroelectric system, why not just add a turbine and start making electricity?" We determined that a hydro plant could generate renewable energy at a fraction of the cost of

Snowmass Microhydro Costs	
Costs (in US$)	
Equipment	$65,610
Turbine and switch gear, structure and foundation	$48,957
Excavation, pipe connection and associated fees	$7,500
Consulting fees	$7,240
Flow meter	$6,000
Electrician	$5,200
Utility interface	$5,000
Shipping	$3,000
Installation and crane	$2,000
Permits	$1,500
Total Costs	**$152,007**
Grants	
CORE/REMP/Ruth Brown Fdn.	−$20,000
OEMC	−$15,000
StEPP	−$10,080
Holy Cross	−$5,000
Town of Snowmass Village	−$5,000
Total Grants	**$55,080**
Grand Total Cost	**$96,927**

Auden Schendler

A Turbine On Every Slope

Think about the possibilities. Hundreds of ski resorts in the US have snowmaking systems. On our four mountains alone, we have half a dozen more good opportunities for hydro. If we had 5 or 10 turbines running, we'd be generating an enormous amount of renewable energy—enough for say, 200 homes—contributing to clean air, stable climate and the long-term sustainability of the ski industry and the town. Any ski resort with a snowmaking system should look into installing a turbine. Inside each of those turbines, you'd find a Pelton wheel, a tool so elegant that it meets Einstein's design criteria that everything should be made as simple as possible, but not simpler. It's a device that has its origins tied to the origins of this town, and now, tied to its future as well.

Project Partners

The Snowmass hydroelectric project is so exciting and forward-looking, and has such broad applicability, that a wide range of partners were interested in providing financial support to help make it happen. Donors included Holy Cross Energy, the utility that buys the electricity and has also covered all grid interface fees (holycross.com); the Colorado Office of Energy Management and Conservation, which supports innovative energy projects all over Colorado (state.co.us/oemc); the Community Office for Resource Efficiency (CORE), which is a national leader in renewable energy and energy efficiency and helped bring a green pricing program to Colorado (aspencore.org); the Renewable Energy Mitigation Program (REMP) from the town of Aspen, which collects fees from new homes

using solar-electric panels. And the return on investment could be as low as seven years.

Convinced that a microhydro system was the best way to generate on-site renewable energy, Snowmass Ski Area built a small powerhouse on Fanny Hill, the beginner slope at the base of the mountain. The building houses a 115 kilowatt turbine attached to a 10 inch steel snowmaking pipe that drains water from a storage pond which is 800 feet farther up the mountain and is fed by West Brush Creek. In 2005, our first complete year of operation, we made some 200,000 kilowatt-hours—enough to power 40 homes—while preventing the emission of 400,000 pounds of carbon dioxide.

that use large amounts of energy (aspencore
.org/NEW_FORMAT/ REMP_new_format
.htm); turbine manufacturer Canyon Hydro,
which discounted its equipment (canyon
hydro.com); the StEPP Foundation (Strategic
Environmental Project Pipeline), whose con-
tribution made Aspen Ski Company (ASC)
the only corporation in state history to receive
money from environmental mitigation funds
(steppfoundation.org); the Ruth Brown Foundation; the town of Snowmass Village (tosv
.com) and Snowmass Water and Sanitation,
which contributed time, space, and technical
support.[2]

Editor's Note: This article first appeared in
Home Power Issue #111 (February/March
2006) and is reprinted by permission of the
author.

From Water to Wire—
Building a Microhydro System

PETER TALBOT

For 500 miles, the remote and storm-battered coast of British Columbia, Canada winds its way north in a torture of craggy cliffs and isolated fjords. It is drenched by the wettest climate in North America and situated at the foot of the ice-covered Coast Mountains.

This wild isolation provides a perfect setting for tapping into the endless supply of energy produced by falling water.

Remote Camp

Tucked among these mountainous wilds, 100 miles north of Vancouver lies the picturesque resort camp of Malibu Landing. Forty-five years ago, a wealthy entrepreneur built the Malibu Club as a private resort for the stars of the California film industry. Boasting all the modern conveniences of the time and situated in a beautiful location, the resort operated for a few brief years before being abandoned due to unpredictable, cool Canadian summers and fierce winter storms. Following the closure, the camp was converted into a summer camp for teenagers and has functioned in that capacity for over 40 years.

Since its early beginnings, this isolated site has been subject to the relentless roar of diesel-powered generators and the high cost of barged-in fuel. It is surrounded by snow-covered mountains up to 8,500 feet high and blessed with steep, flowing creeks. The site was a natural for a microhydro power plant, yet in all these years one had never been developed.

I had been visiting the area and volunteering at the camp for a number of years and saw the potential for a development that could reduce their dependence on diesel fuel. For most of the winter, a thin waterfall cascades over cliffs 1,000 feet above the camp. Though dry for most of the summer, this was a potential source of hydro power for the winter months.

Since the camp is closed in the winter, the power requirement for the year-round caretaker is small, averaging under ten kilowatts and might just be handled by a small hydro plant fed from this seasonal flow. A decision was made to conduct a rough survey of the terrain and then collect stream flow data over the course of the following winter. If the flow proved to be sufficient, we would begin construction the following summer.

The Survey

One of the first steps in the design of a hydro plant is to determine if there is sufficient flow available to make the project worthwhile. Fortunately, the wet winter season corresponded with the demand that would be placed on the system, and long-term casual observations suggested that there would be adequate flow for most of the winter. The caretaker had been keeping an unofficial visual record for almost ten years and could compare the estimated flow on any given day with seasonal norms. This proved to be a great advantage when we installed an accurate measuring device at the falls, since we could then compare actual flows with past observations.

The survey team at the base of the falls, ready to measure total head.

Peter Talbot

Measuring Head

The second key ingredient to a successful hydro project is the total available change in elevation over which the water can develop pressure in the pipeline. We first measured this *head*, or elevation drop, by means of a sensitive altimeter, and then with a handheld clinometer level and a 15 foot survey rod.

The route the pipeline would take was more or less obvious, so we followed this as we carefully took each reading off the rod. As we leapfrogged up the hill, the exact elevation was marked on prominent landmarks as a permanent record. The use of the rod and level gave considerable accuracy over the distance, which traverses some really rough terrain. Two elevation surveys were made to check for error and the results tied within a foot — close enough considering the method used.

When all the surveyed elevation steps were added up, the total to the base of the falls came to 639 feet above the proposed powerhouse floor. The altimeter reading agreed within ten feet and provided a good check against any gross errors. This elevation is on the high side for the typical microhydro installation, but it allowed us some margin for locating an open filter box and starting the pressure penstock.

Increasing height raises the operating pressure and hence the power output. However, it also causes the turbine to spin faster, increasing with the square root of the height. This affects the turbine diameter used, the desired output frequency and the pressure rating of the piping.

Sizing Pipe

To measure the overall distance, we used a 100 foot survey tape and again marked the distance along the route. The total came to 2,200 feet, of which about 2,000 feet would form the pressure penstock. Determining the distance was much easier than measuring the exact head, but it too had to be done carefully, since we planned to use pre-cut steel pipe lengths in the lower section.

We planned to use high-density polyethylene pipe (HDPE) for most of the pipeline. Since the static water pressure would be increasing as the pipeline descended the slope, we had to decide where we would change to the next greater pressure-rated pipe. We did this by dividing the slope into six pressure zones and selecting the appropriate pipe thickness for each zone. This HDPE pipe is extruded in various thicknesses. Often the pipe is rated by a series number, giving its safe sustained working pressure. Another common system rates the pipe by its dimension ratio (DR), which compares the pipe's wall thickness to its diameter.

We planned to use DR26 in the low pressure section, which is the same as series 60, all the way up to DR9, which is equivalent to series 200. Beyond that, the wall thickness increased enough to significantly reduce the inside diameter. This would cause the water flow velocity to increase, resulting in greater friction and hence losses, so a strong, thin-walled steel pipe became a better choice and cost less.

Determining the Required Flow

Since the survey was done in summer when there was just a trickle of water flowing, we didn't have the actual flow data. As a result, we couldn't calculate the exact power output, efficiency and payback time. However, having a fixed budget to work with and knowing the head, distance, penstock profile and power requirement, it was possible to design a system based on a minimum anticipated winter flow. Calculations showed that half a cubic foot per second, or about 225 US gallons per minute, over a net head of 500 feet would produce an output of 12 kilowatts and make the project well worthwhile. A simple formula to estimate electrical power produced from falling water in an AC hydro plant of this size is as follows:

Power in kilowatts equals Q times H divided by 11.8 times N

where Q is flow in cubic feet per second, H is head in feet and N is overall efficiency (typically 60% (0.6) in a small, well-designed system).

Another version of the power output formula is:

Power in watts equals net head in feet times flow in US gallons per minute divided by 9

This formula already takes efficiency into consideration. For this site, the result was 500 feet times 225 US gallons per minute divided by 9 which equals 12,500 watts (or 12.5 kilowatts).

Measuring the Actual Flow

In order to get an accurate record of the flow profile over the winter, we constructed a wooden tank equipped with a V-notch weir and placed it below and to the side of the falls. A length of six-inch diameter plastic pipe was secured in the channel to catch the majority of the runoff and direct it into the box. The depth

V-Notch Weir Chart

Notch depth (inches)	1	1.5	2	2.5	3	3.5	4	4.5	5	5.5	6	6.5	7	7.5
US gallons per minute	2.4	6	12.6	21.6	34.2	51.6	72	96	126	162	198	240	288	342

Peter Talbot

of the water flowing through the calibrated V-notch weir gave an accurate measure of the flow available.

Details on building various weirs are outlined in most textbooks dealing with fluid flow. These are available in many libraries. We used a 90° V-notch weir cut out of a piece of sheet metal. Figure 4 shows the flow in gallons per minute per inch of depth through a small V-notch weir.

A sensitive water-level monitor was installed in the box, coupled to a radio transmitter which would relay the flow conditions down to the camp every few hours. A modified receiver and some additional electronics showed the level on a numeric display, which was read and recorded by the caretaker. He could then compare this accurate flow reading to what he observed flowing over the falls and relate this to his ten years of casual observations. As the long, wet winter set in, it soon became clear that there would be more than enough flow to make the project viable, so we began to design the system.

The intake box is used for filtering and settling of debris. The V-notch was used for determining flow during system planning.

Peter Talbot

Several years of use has proven the intake basin's covering of rock a worthy armor and a coarse filter.

Peter Talbot

Shopping List

Once we had the approvals to build the project and had established a preliminary budget of Can$15,000, the next phase was to order the necessary hardware. We were fortunate in that most of the suppliers were willing to give us jobber prices, since Malibu operates as a non-profit organization.

Since we had done an accurate survey, we could order the pipe to the exact length and pressure rating that we required. We went to the suppliers before ordering the materials to check out the quality of the steel pipe and to be sure that we would be able to handle the weight during construction. Pipe lengths of 20 feet weighed 180 pounds and would have to be carried by hand over very rough terrain.

The four-inch diameter polyethylene plastic pipe was ordered in 40 foot lengths. The pressure ratings varied from 60 pounds up to 200 pounds with a safety margin of 25%. Transporting the pipe was expensive since it required a 40 foot truck to get it to a suitable waterfront dock where a landing barge could be loaded. The long lengths did, however, cut down on the number of joints we had to make.

One of the advantages of using polyethylene pipe over PVC is that the working pressure can be close to the pressure rating of the pipe itself. This is due in part to the elasticity of the plastic used, which will absorb the shock wave (water hammer) generated if the water flow is forced to change velocity abruptly. This effect causes a momentary pressure rise which travels up the pipe and has the potential to do permanent damage, even bursting a more rigid pipe.

To further reduce possible damage to the pipe when shutting off the flow, we obtained a slow-acting four-inch gate valve. This was picked up at a scrapyard for Can$50! With a pressure rating of 500 pounds, this valve would have cost many times that if purchased new.

Pelton Wheel

The high head and relatively low flow rate of our site would be best handled by a Pelton type of turbine. Since our operating head would be somewhere between 500 and 550 feet and we wanted the rotational speed to be 1,800 revolutions per minute — suitable for direct coupling to a generator — we needed a turbine with a diameter of approximately ten inches. When under load, this diameter wheel would rotate

John Smocyzk, a regular volunteer at Malibu, shows off the fusion welding equipment for the polyethylene pipe.

Peter Talbot

at the correct speed and the direct coupling would afford the maximum efficiency.

We looked at three different turbines and got firm quotes. Each machine had its own merits, and costs were roughly equal. We settled on a unit made by Dependable Turbines, a local manufacturer, because of their proximity to, and familiarity with, our site. They also had a turbine runner with the correct pitch diameter and bucket size to exactly match our site characteristics. The turbine was ordered as a package, together with a 14 kilowatt, three phase Lima brand generator.

Intake

Intakes are usually the most difficult aspect to design on a microhydro project. Seasonal variations in flow can range from a trickle in late summer to a raging torrent in winter. On the steep mountainous terrain of the west coast, many a concrete intake structure has vanished following a heavy downpour.

With this in mind, we thought about ways we could minimize the construction required and work with the natural form of the land. It was obvious that ice and rock falling from the frozen lip of the falls high overhead would destroy any structure we built.

What was needed was an intake that was formed as much as possible from the bedrock buried beneath the boulders and gravel below the falls. Following some excavation, we were able to take advantage of the sloping granite bedrock down the hill from the base of the falls and out of the direct line of falling ice and rock. We built a low wall of reinforced concrete there to divert the flow into a small pool, enabling us to pick up even the smallest flows. The pool and wall were then backfilled with large rocks. Falling rock and ice would then pass over the low wall, leaving it undamaged.

From the pool at the 600 foot elevation, we ran four-inch plastic pipe for 200 feet across and down to a level spot at the 550 foot elevation. We moved the five-foot-long wooden box that was used to measure the flow to this spot. Then we equipped the box with three sizes of filter screens and a valve in the bottom to allow for the flushing of any sand and gravel. Excess water passes through a narrow one-inch slot cut into the top 12 inches of the tank which forms the overflow. This replaced the V-notch weir and increased sensitivity for the level sensor. A pressure transducer and microprocessor circuit relays the level of overflow to various locations in camp by a radio link and phone wires. This allows the operator to monitor the flow and to throttle back on the water passing through the turbine as the falls dry up. When there isn't enough water to make it worth running the turbine, the operator can switch over to diesel. From the filter box, the pressure penstock runs 2,200 feet down to the powerhouse, dropping 550 feet.

Laying Pipe

The great advantage of polyethylene plastic pipe is that it is almost indestructible. It is not affected by UV exposure, can be squashed nearly flat and recover and can freeze solid under pressure and not split. The major disadvantage is that it cannot be glued, but must be either fusion welded or connected with expensive hugger clamps. We opted to rent the welder and join the 40 foot lengths into long sections at the bottom of the hill where there was the necessary 1,500 watts of 117 volt power to run fusion welding equipment. It was quite a sight to see the first section of pipe stretch for 400 feet down the dock and float halfway across the bay as more sections were welded on!

The welding process is really a form of hot fusion melting. This involves placing the pipe ends in a special holding jig and squaring the ends with a motorized cutter which is inserted between the pipe ends. The pipes are brought together in the jig and contact the cutting wheel which planes off a bit of plastic. The cutter is then removed, and a flat heated plate inserted. The pipe ends are lightly pressed against the hot plate for a minute or so to soften the plastic. Then the plate is removed and the pipe ends are brought together under light pressure. A bead of plastic forms as the melted plastic fuses together. After cooling for five minutes, the joint is complete and is said to be stronger than the rest of the pipe. Despite some very rough handling, we have never had a leak.

When ready, we got another 20 volunteer grunts to help haul the pipe up the hill following a carefully surveyed path. This was a lot of fun, but also an amazing amount of work. We were fortunate to have the willing bodies.

Most of the plastic pipe was laid directly on the ground and secured to solid trees and rock anchors with ½ inch white nylon rope. We found that yellow poly rope would not last long in the sun. Pipe destined for the lower sections of the route was much heavier, so we

Top: Floating 400 feet of poly pipe across the bay to the base of the hill.

Bottom: A crew of up to 25 volunteers haul 400 foot sections of polyethylene pipe up 550 vertical feet to the turbine.

A hugger clamp joins poly pipe to steel pipe.

Peter Talbot

Down through the trees, the bottom sections of steel pipe reach for the powerhouse.

Peter Talbot

welded these into lengths of 160 feet, intending to join the long sections with hugger clamps. These clamps are made of two halves that bolt together and compress sharp ridges into the pipe wall. A rubber gasket makes them watertight. Although expensive, with enough of these clamps the entire penstock installation could have been done by two people.

We soon found that our small one-kilowatt Honda generator would run the welder if we momentarily unplugged the hot plate when we needed to use the cutter. So we decided to haul the equipment up the rough route and weld the plastic pipe into one 1,700-foot-long piece. This gave us a slightly smoother pipeline, and it allowed us to keep the expensive hugger clamps for future repairs to the line.

Steel Section

The 20 foot lengths of steel pipe were muscled up the hill one piece at time by three bush apes and connected together by Victaulic clamps. This is a two piece cast fitting that is bolted together and grips into grooves cut into the pipe ends. A rubber gasket prevents any leaks. This method of coupling allows a few degrees of flex at each joint while avoiding the need for an arc welder.

Each 20 foot length of steel pipe weighed 180 pounds, and we put in 550 feet of it. As the line was extended, we supported it on rock and timber cribbing at regular intervals. Half-inch wire cable was wrapped around the pipeline just below a coupling, then clamped together forming a small loop. We attached the cable to one-inch diameter rods drilled into rock outcrops and tensioned it using a come-along (hand winch).

Bends were kept to a minimum, and where necessary we used short 22.5° pre-formed sections. By anchor planning the route carefully and aiming for solid anchor points, we were able to obtain a perfect fit with just four bends. Our main anchors and thrust blocks were drilled into solid bedrock. We used a portable electric rock drill which worked very well. It was able to cut a one-inch diameter hole, four inches deep, in under five minutes.

Just in front of the powerhouse, the penstock crossed a small bay. Here we built up log scaffolding to hold the pipe as we maneuvered it into the most direct route while correcting the slope so it would be self draining. Once the position was established, we waited for low tide, then placed forms directly below the pipeline. Pilings were set vertically in the forms, and the forms were filled with underwater-setting concrete. The thrust block at the powerhouse keeps the tremendous weight of pipe and water from sliding downhill and crashing through the building.

After three days, the penstock was slid over on the pilings and secured, and all the scaffolding was removed. Once the penstock was secured in place and the main valve attached, we began the pressure test by slowly filling the pipe from the trickle coming over the falls. It sagged in places and pulled against the cable anchors, but there were no leaks. When it was full, the static pressure read 239 pounds, which was within a pound of what had been

Top: Pipe anchors were drilled into solid rock.

Middle: The steel pipe comes out of the woods and across the bay to the powerhouse.

Bottom: Volunteers Dave Wheeler and John Smoczyk built scaffolding to support the 180 pound sections of pipe.

Peter Talbot

Peter Talbot

Peter Talbot

calculated. A static pressure penstock will develop 0.433 pounds of pressure for every foot of vertical drop. In our case, the measured 550 feet of head should then give 238.1 pounds per square inch (550 feet times 0.433 pounds per foot equals 238.1 pounds per square inch).

Powerhouse

The site for the powerhouse was selected to minimize the overall penstock length and the number of pipe bends required. We wanted easy access and a location safe from ocean swell and any freak high tides. The machinery and related controls required a space of about 9 by 11 feet. This would give access to all sides of the turbine for maintenance and installation, which later proved invaluable.

In order to get a solid anchor, the bedrock was cleaned with a fire hose and then drilled for steel reinforcing bar. A wood frame was built on three sides of the sloping bedrock, and backfilled with concrete and broken rock. Mechanical drawings of the turbine showed how large to make the *tailrace*, or discharge pit, so this was formed with a bit more framing. A notch for the generator power conduit and other control and monitoring wires was formed before the final surface was smoothed.

Installing the turbine was simply a matter of placing it over the tailrace pit and drilling the concrete to line up with the holes in the steel flange forming the turbine base. The generator bolted directly to the same base and required a few shims for correct alignment. A semi-flexible coupling joined the two-inch turbine shaft to the generator shaft.

Left: The thrust block at the powerhouse keeps the tremendous weight of pipe and water from sliding downhill and crashing through the building.

Middle: Camp caretaker Frank Poirier, on the powerhouse concrete foundation, with framing for the tailrace visible. The building was built around the turbine and generator.

Right: The generator and turbine visible in the powerhouse. The tailrace dumps out the side of the foundation.

Peter Talbot

The pressure penstock terminated at the main valve just inside the powerhouse walls. Right outside, the penstock was securely anchored to a huge rock outcropping. This formed the final thrust block, and restrained the downward force the weight of water and pipe imposed against the valve body. Over the four-inch diameter, the total force was close to 3,000 pounds, so a solid anchor was essential.

From the valve, we connected the intake manifold to the nozzle flanges which were part of the turbine housing. A couple of four-inch sections joined by Victaulic clamps were added between the valve and the main thrust block to give a little flexibility and expansion relief. This is important and prevents possible cracking as expansion and contraction vary the dimensions of the steel.

The powerhouse was framed up and the roof built over the installed machinery. A requirement was that it had to blend in with the other old log and cedar building on the site.

We were fortunate to have a skilled carpenter who was familiar with building to exact specifications.

Controls — How It Works

The Pelton turbine is equipped with two nozzles, each with a maximum diameter of 0.5 inches. One of these is equipped with a spear control (similar to a needle valve in a carburetor, but much larger). This allows the flow rate to vary. This is necessary when the flow is lower than what a single ½ inch nozzle would require. With this adjustable spear, we can run the turbine with very little water and still get useable power.

The generator was chosen for the best efficiency rating at the mid-range of our power demand. When there is too little flow, the die-

Left: The powerhouse blends in with the forest and the traditional buildings on the site. The penstock enters the rear of the building.

Right: The controls and metering on the powerhouse wall.

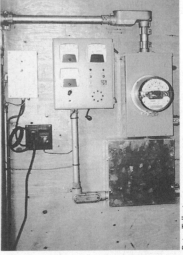

Peter Talbot

sel is used. In times of high flow, there is more than enough water, so efficiency is not as important. This same principle can apply to any small run-of-the-river system.

Most synchronous generators come equipped with 12 output leads. They can be hooked up to produce single phase or three phase current. This usually depends on the application. A typical home situation would most likely require single phase power, at 120 and 240 volts.

Larger installations and any site with big industrial motors usually require three phase power. This was the situation we were faced with. The 125 kilowatt diesels used in summer fed the camp's three phase grid, so to avoid very complex rewiring, we wired the hydro generator accordingly. The major load was the

The 14 kilowatt Lima generator is direct coupled to the Dependable Turbines Pelton runner.

Peter Talbot

caretaker's house, and this was wired like any conventional home, drawing juice from only two of the three phases. Other loads could be connected to the third phase to maintain a better balance on the generator. Three phase generators can be damaged if they are run with all the load on just two of the three phases.

Sixty Hertz Governor and Load Dump

The generator is directly coupled to the turbine through a semi-flexible coupling. So in order to produce standard 60 hertz, the turbine must spin at exactly 1,800 revolutions per minute. This is accomplished by using a Thomson and Howe electronic governor, which works by keeping a precise but constantly varying load on the generator. In essence, it *puts the brakes* on the generator and turbine if it deviates from 60 hertz.

The governor works by sensing the generated power line frequency and comparing this nominal 60 hertz to a crystal reference. An internal microprocessor then controls the phase firing angle of high power triacs which shunt excess power to low priority, but useful, dump loads.

These loads do not necessarily see the full sine wave generated since they are being fed with rapidly switching and varying width pulses. Because of this, only purely resistive loads can be used; motors or electronics would soon self-destruct. We used baseboard heaters located in a large woodworking shop. Immersion elements in hot water tanks are another useful dump load.

Frequency stability is excellent with this method of control, and it avoids the much more complex method of mechanically controlling the flow of water to precisely match the electrical load. This was traditionally done with centrifugal weights acting on an oil-based servo control, which in turn controlled a deflector in front of the nozzle or a spear valve.

Protection: Shaft Speed and Frequency

The frequency of the system is monitored by two independent systems. Should the generator begin to slow down due to excess load, or possibly overspeed due to insufficient dump load or a broken power line, the protection circuitry will sense the condition and shut the machine off. This is accomplished by optically sensing the shaft speed as well as line frequency and voltage. The frequency limits are user-adjustable.

Without this protection, motors and transformers would be subject to lower than normal line frequency which can cause damage. As the generator slows, the frequency falls in direct proportion to the revolutions per minute, while the generator's voltage regulator tries to hold the voltage constant. This can cause large currents to flow in the regulator and field windings as the regulator tries to maintain the output voltage. Generally, resistive loads like incandescent lighting and heating elements are not damaged by low voltage or frequency, but reactive loads, such as devices with windings like motors and transformers, are at risk.

These frequency, speed and voltage sensor outputs are connected to a weighted mechanical jet deflector which will divert the water away from the turbine runner. A magnetic latch holds the deflector in the open position in the absence of an alarm. An adjustable time delay will release the latch in the presence of an alarm condition, shutting the system down. This requires a manual restart which is a bit awkward if it happens in the middle of the night. But the consequences of the turbine lugging or running away at high speed can be very bad.

Metering

Voltage and current are displayed on a home-brew metering panel, together with alarm status, water level indication and shaft revolutions per minute. The water level is also displayed at other locations in the camp, and the displays are equipped with an adjustable low water alarm set point. This keeps the operator informed of the flow situation up the mountain and provides advance warning of when to switch over to the diesel generator.

We also installed a three phase kilowatt-hour meter to monitor the total energy produced. This added feature has enabled us to keep track of the savings in diesel operating costs and to determine how the project payback is proceeding. It is really satisfying to see the meter whiz around and to know that the small creek is powering all our needs. The best part is that for the first time in 40-odd years, there is complete silence throughout the camp, yet all the lights are on!

Breakers and Switch

A 60 ampere fused disconnect feeds into 300 feet of #4 Teck cable (outdoor armored cable) which runs from the hydro site to the diesel powerhouse. The hydro output can then be fed into the main bus system and distributed throughout the camp as required. We had to install a triple-pole double-throw transfer switch so either the hydro or a small 15 kilowatt diesel generator could feed into the camp grid. One, but never both of these, is always supplying power.

The transfer switch then feeds a 60 amp circuit breaker which in turn feeds into the camp's grid. This last panel has two keys which must be turned before it can be put on line. Both of the two main diesel generators (125 kilowatt and 113 kilowatt) also feed into the grid through separate breaker panels. The same key must be used in both of these panels before they can be switched on. This eliminates any danger of backfeeding one generator into another.

Life With Hydro

As the winter rains returned, the falls once again began to pick up force. On a rainy day in late October, the telemetry system indicated a flow through the catchment weir sufficient to test the system. The penstock pressure gauge read 239 pounds under the static head of 550 feet.

Once the pipeline was purged of debris, the spear valve was cracked open, and the Pelton wheel immediately started to rev up. At first we set it to produce just a few amps, letting the governor dump load absorb the output. The effort we had made to align the shafts with the correct thickness of shims during the installation phase was rewarded by quiet operation with virtually no vibration. Once it checked out, we opened the spear valve, and the output quickly increased to 20 amps per phase. As predicted, we were getting close to six kilowatts using one nozzle!

Other than the silent operation, there is no way to tell that the camp is running on renewable energy. Under wet conditions it will run for weeks without stopping. We were accustomed to shutting the diesel down every day and adding oil, so this took some getting used to!

A fixed amount of water flowing through the turbine sets the limit on power production. Unlike the diesel, there is no throttle which will automatically open up as the load increases. To attempt to draw more energy out than is being supplied by the water jets will result in the system slowing down. Drawing even a few extra watts slows the shaft speed and hence frequency, and the turbine will shut down.

A system that will trip itself off on overload is a minor inconvenience, but is something one learns to live with. The protection it affords is definitely worthwhile. It doesn't take long to approximate the electrical load on the system. If a load larger than the governor reserve is switched on, the line frequency begins to fall. If you are quick, you can switch it off again and the turbine will recover. Over time, the kilowatt-hour meter began counting up in the thousands of kilowatt-hours. It was obvious that the payback would take just a few winters at this rate!

Lessons

Two factors which produce the only notable trouble are the intake clogging up and the variable flow of the water source. The clogging can be minimized by using effective screening (see the article on Coanda screens in *Home Power* Issue #71). We have not tried this approach yet, but rely instead on several large wire mesh baskets and regular cleaning by hand. The problem is only bad in late fall; throughout the winter there is little debris in the water.

Times of low flow still produce a useful output which provides additional heat even when the small diesel is running. In fact, we can leave the turbine unattended under this condition. The plant will keep on running, feeding into the dump loads, producing heat for the workshop. When it gets down to the last few hundred watts, it will quickly shut itself off when the water probe signals that the intake box is low on water. At this point we close the valve so the penstock doesn't drain. The only exception is if a hard freeze is expected. Under this condition, the line is drained.

One big lesson we learned quickly was that it is one thing to design a system based on summer conditions, and quite another to implement it and expect it to withstand the ravages of a winter storm. Rock fall and sheets of ice falling from high above will destroy just about any structure. We had to adjust our intake piping several times to prevent it from being swept away. We finally buried it, and it has been safe since then.

The catchment weir has been a big success. There is evidence that some really large rocks rolled over it, and it has been buried under a mound of ice several feet thick. The only minor trouble is the four-inch outlet pipe clogging with gravel and vegetation. We plan to replace this with a short length of six-inch pipe and screen out the major debris with a coarse screen, followed by a Coanda screen.

Work or Play?

By far the hardest part of this project was the installation of the 2,200-foot-long penstock. We chose to haul long sections of pipe up the hill by hand, and at times we had 25 bodies spaced along the section, all straining away.

When we found that the fusion welder could be run off the small generator, we packed it up the hill.

It took a crew of four guys to pack all the welding equipment, and several more to assist in aligning the pipe prior to fusing the ends. It's not backbreaking work, but it does demand a coordinated effort. Despite the complexity of working with this polyethylene pipe over PVC pipe, I would do the same thing again. Poly pipe is so amazingly strong and flexible; it's the only material that could stand up in our situation.

The steel section went together surprisingly quickly; it took just two days to place all 550 feet. Having a ready supply of blocking material and having pre-drilled the anchor points allowed us to connect the sections as fast as they could be carried up the hill.

The scaffolding we had set up over part of the bay enabled a crew of just three to connect the sections. Constructing the scaffolding took extra time, but it was worth the effort. Working with heavy pipe overhead is risky enough, so it was worth taking the time to do it safely. Having a volunteer labor force available at the camp was the biggest saving. Without this, the project would have taken much longer, and the construction cost would have been considerably higher.

Efficiency

At the maximum flow of half a cubic foot per second, we are able to produce 35 amps per phase. This works out to 12.6 kilowatts, spread between our main loads and the governor's resistive dump load. With a flow of 225 gallons per minute over the falls, and a gross static head of 550 feet, there are 23 kilowatts of

Malibu Club Hydro System

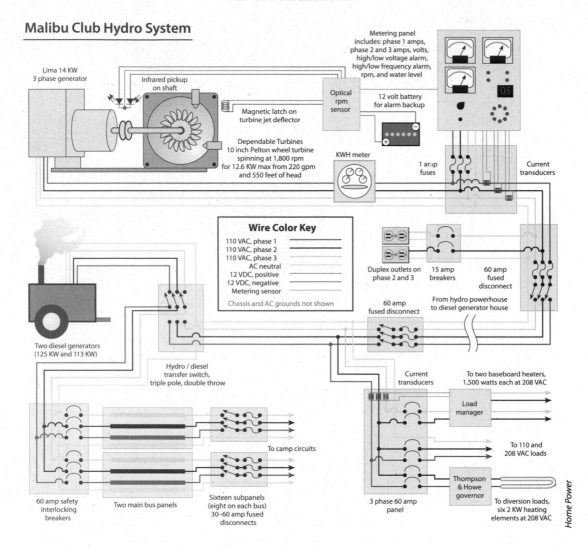

Lima 14 KW 3 phase generator

Infrared pickup on shaft

Magnetic latch on turbine jet deflector

Optical rpm sensor

Metering panel includes: phase 1 amps, phase 2 and 3 amps, volts, high/low voltage alarm, high/low frequency alarm, rpm, and water level

12 volt battery for alarm backup

Dependable Turbines 10 inch Pelton wheel turbine spinning at 1,800 rpm for 12.6 KW max from 220 gpm and 550 feet of head

KWH meter

1 amp fuses

Current transducers

Wire Color Key

110 VAC, phase 1
110 VAC, phase 2
110 VAC, phase 3
AC neutral
12 VDC, positive
12 VDC, negative
Metering sensor

Chassis and AC grounds not shown

Duplex outlets on phase 2 and 3

15 amp breakers

60 amp fused disconnect

60 amp fused disconnect

From hydro powerhouse to diesel generator house

Two diesel generators (125 KW and 113 KW)

Hydro / diesel transfer switch, triple pole, double throw

Current transducers

To two baseboard heaters, 1,500 watts each at 208 VAC

Load manager

To 110 and 208 VAC loads

To camp circuits

Thompson & Howe governor

60 amp safety interlocking breakers

Two main bus panels

Sixteen subpanels (eight on each bus) 30–60 amp fused disconnects

3 phase 60 amp panel

To diversion loads, six 2 KW heating elements at 208 VAC

Home Power

potential energy available. Our 12.6 kilowatts represents about 55% of that total.

Hydro

An efficiency figure of 60% is about average for a small system such as this. Our turbine is rated at 76%, and the generator 79%. We lose about 10% of the gross head due to friction in the penstock at full output. Totalling this

(79% times 76% times 90%), we have 54%, and 54% times 23 kilowatts equals 12.4 kilowatts, roughly our measured output.

On average, the system is set with only the adjustable nozzle open. This will produce just under seven kilowatts. The reduced flow velocity results in slightly less pipe friction. This in turn results in higher net pressure at the turbine, and the more efficient spear nozzle appears to account for the increase in overall efficiency under this condition.

Payback

The 15 kilowatt diesel generator would go through an average of two gallons of fuel per hour. At Can$2 per gallon, the cost to run the diesel works out to Can$4 per hour or Can$96 per day. That comes out to 27 cents a kilowatt-hour for fuel costs only. We used this figure to calculate the payback time of the hydro plant. On average, we produce six kilowatts and can run for about 100 days a year. If we price the hydro power at the same rate as diesel-produced power, our hydro is earning Can$39 per day (6 kilowatt-hours times $0.27 per kilowatt-hour equals Can$1.62 per hour equals Can$39 a day). That's Can$3,900 per season, so it will pay for itself in just under four years. Not a bad investment!

As mentioned earlier, we were able to keep the total project cost down by doing some scrounging and by purchasing new equipment at a slight discount. Other items were available on site (such as building materials), and all the labor was donated. The electronic water level sensor and optical frequency control were built at cost.

With the great success of this project, we are now planning to construct a larger plant on a year-round creek two miles from the camp. This would take care of the needs of the entire camp throughout the year, and would result in significant cost savings. On behalf of the Malibu Club, I wish to extend my thanks to all those volunteers who helped make this project a reality. In particular, thanks to Ron Kinders, Malibu's representative. Without his continual dedication and assistance in

some very demanding conditions, this project would never have gone ahead.[1]

Editor's Note: This article first appeared in *Home Power* Issue #76 (April/May 2000) and is reprinted with permission of the author.

Malibu Club System Costs	
Turbine and generator	8,500
1,800 feet of plastic pipe	2,400
Governor	1,145
350 feet of four-inch steel pipe	740
8 hugger clamps	350
Switch gear (some reconditioned)	325
Metering and level sensing panel	300
Welder rental for five days	250
Dump loads	250
20 Victaulic clamps	240
Rock anchors and cable	225
Additional wire	145
Intake box and screen	100
Four-inch gate valve (scrap)	50
Pressure gauge	42
Concrete dam	20
Subtotal	**$15,082**
Other	
Powerhouse*	1,300
600 feet of #4, four-conductor Teck wire**	1,200
Total	**$17,582**

Note: all figures in Canadian dollars.
* Built with materials on hand, not included in original budget.
** Teck cable was a later addition.

Peter Talbot

Hydro Power High
in the Canadian Rockies

PAUL CRAIG AND ROBERT MATHEWS

The Canadian Rockies offer some of the world's most spectacular outdoor experiences: deep powder skiing, alpine hiking and incredible glacier views. The Purcell Lodge is deep in these Canadian Rockies, near the town of Golden, British Columbia, four miles beyond the nearest (logging) road. Every piece of equipment, all supplies and the guests are brought in by helicopter. Out here everyone greatly appreciates the light, heat and appliances made possible by the lodge's 12 kilowatt hydroelectric system.

A Canadian Classic Hydro System

The electricity generation system uses what has become, in Canada, a *classic* hydro setup with a high head, small pipe and Pelton wheel turbine. Turbine speed regulation is accomplished by electrically loading the generator to maintain frequency. The location was selected to provide year-round water. Fortunately an insulating cover layer of snow always arrives before ground-freezing weather. Even though the snow season can last six months, there have been no problems with frozen pipes.

The intake weir (head pond, see bottom photo, pg 28) is a concrete wall in a largely

bedrock location that provides a small impoundment basin. The pond stills the flow, allowing heavy debris to settle and light debris to float. The intake pipe is submerged at half-depth and screened. The pond also contains a submersible pump for domestic water.

The penstock is 1,440 feet of four-inch diameter solvent-welded PVC pipe, with pressure rating increasing from 63 pounds per

The Purcell Lodge in winter with Copperstain Mountain in the background. Winter snowfall here is about 55 feet. Without site-produced electricity, modern living in such a remote location would be impossible.

Paul Lesson

Top: The eight-inch hydroelectric Pelton turbine (painted silver), flywheel (painted red) and 12 kilowatt generator (painted yellow). The Fidelity brushless generator produces 60 hertz, 120/240 volt single phase alternating current and is regulated by a Thompson and Howe controller.

Bottom: Intake weir and top of 1,440 foot long penstock which feeds water into the turbine. Total head is 315 feet, and maximum flow is 220 US gallons per minute.

Bob Mathews

square inch at the intake to 160 pounds per square inch at the turbine. Total head is 315 feet, and maximum flow is 220 US gallons per minute. The penstock is buried 18 inches and anchored with concrete and bolts at critical points.

The eight-inch diameter Pelton turbine was built by IPD of Montana. When spinning at 1,800 revolutions per minute, the wheel moves at 46% of the jet speed at the point of contact. (At 41% to 47% of jet speed, Pelton wheels are most efficient.) The main nozzle is manually adjustable through a spear-type valve. Maximum nozzle diameter is $^{13}/_{16}$ inch.

The generator is a Fidelity brushless design rated at 12 kilowatts at 1,800 revolutions per minute. Maximum power output was initially limited by the available flow to seven kilowatts. During the fall of 1992, the water supply system was modified to provide the full 12 kilowatt output. Friction loss in the penstock is about 8% at seven kilowatts, and the turbine-generator converts the water energy reaching it with 56% efficiency. Output is 120/240 volt, 60 hertz single phase AC electricity. The generator is direct-driven from the Pelton wheel. A flywheel maintains speed under high starting loads from induction motors and provides general stabilization.

To decrease energy loss and save wire costs, the voltage is transformed up to 600 volts for the 1,750 foot run from the powerhouse to the lodge. At the lodge, a second transformer provides 120/240 volt output. Two #6 AWG RWU copper conductors are mechanically protected by a one-inch poly pipe and placed under the four-inch PVC penstock to provide further mechanical protection. Transmission power loss is 2.5% at seven kilowatt output.

Control by Prioritized Loads

Primary load control is perhaps the most interesting part of this system. The Lodge uses an Electronic Load Control Governor (ELCG) manufactured by Thomson and Howe Energy Systems. It's easiest to understand this regulator by contrasting it to more traditional control approaches. Solar systems are usually limited by energy. Design focuses on minimizing load and on turning off loads when not needed. Hydro plants are traditionally controlled by regulating water flow as load varies. Not so in the Purcell environment. Here as in many modern microhydro situations, the water runs whether used for electricity generation or not. Since water is not trapped in a dam, the ecology of the stream is less impacted by this type of hydro system. But, if electricity is not generated, the energy in the falling water is lost.

This makes possible a very different type of regulation based on using all available water, while switching on and off priority loads to maintain a constant overall load. The more the electrical loading, the slower the turbine and generator run. Since the generator frequency is directly proportional to rotational speed, control of frequency automatically provides speed regulation. A rise in frequency means that more load is needed to slow down the generator. A drop in frequency means load must be shed.

The lodge load is broken down into a number of circuits. The highest priority circuits, especially those that are safety related and those (such as lights) needed to maintain guest satisfaction, are wired directly to the generator service panel and are not under governor control at all. Lower priority services are connected to the ELCG. Eight circuits are currently in use. These are ranked by priority, with sewage aeration and water pumping followed by refrigerators, freezers, furnace fans and the dish and clothes washers.

Coarse and Fine Control

Coarse regulation is provided by load shedding. For example, if the water system — a high priority function — turns on, a refrigerator or freezer might be temporarily shut off. Coarse control necessarily leads to large and fast changes in system loading. Without additional control this would lead to unacceptably large frequency (and hence turbine speed) swings. Here's where the ELCG proves its merit. Load swings are virtually eliminated by continuous, rapid control of resistive loads such as baseboard heaters and hot water tank heating elements.

For this fine control, the ELCG uses triac regulators to smoothly change the power delivered to resistive loads, increasing as other load drops and decreasing as other load picks up. If the ELCG senses the frequency is rising, it knows that a load is being shed (perhaps someone turned off a light) and increases power to the resistive load. If frequency drops, the ELCG smoothly sheds resistive load. In operation the system is almost unnoticeable. The only indication is occasional slight dimming of lights when a large motor starts up. But of course this occurs with utility power too.

If the load increase is too great to handle with the resistive load alone, the ELCG throws a relay to drop the lowest priority load connected. As load decreases again, the highest priority non-connected load is reconnected. There is a special circuit to keep track of and slowly correct for short-term excursions from

System Diagram

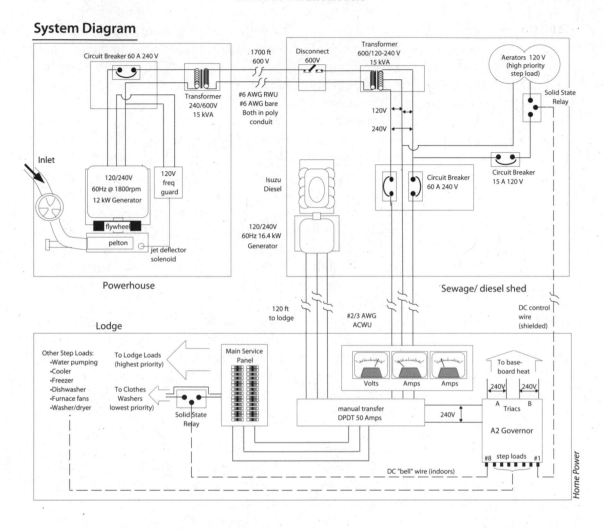

60 hertz due to extreme conditions, so that clocks will keep proper time. Maximum frequency correction is kept to 0.1 hertz.

The wave form from the generator is excellent for all purposes. However, the switching triacs introduce considerable waveform distortion in the power going to their loads. Resistive balancing loads, such as hot water heaters and baseboard heaters, are used which are indifferent to waveform. It is important to assure that the system always has enough load

available to maintain frequency. To accomplish this, priority loads and the resistive regulating loads must be connected at all times.

Any regulation system based on the concept of loading is vulnerable to open circuits, which would lead to system runaway. Failsafe emergency protection is required. This is located adjacent to the generator. A mechanically interconnected water jet deflector safety system is actuated by a weighted lever. The turbine is shut down by a frequency guard sensor if frequency deviations become too wide (typically outside the range 53–67 hertz) for too long. The weighted control lever is held up by a normally energized solenoid which releases on power failure.

Heat, Biodegradable Soap and Solar Radios

Although the building is heavily insulated, auxiliary heat is needed in winter. Since the available water flow doesn't provide enough energy to heat the building under extreme conditions, propane is used for backup. Because the lodge is above timberline, firewood must be helicoptered in. Sewage is handled with a small biotreater plant. Treated waste goes to a carefully monitored leach field. To minimize loads on the facility, biodegradable products are used exclusively. Guests are asked to use the biodegradable soap and shampoo provided rather than any they may have brought.

At Purcell Lodge reliable communications can mean the difference between life and death. A radio repeater on a nearby mountain provides complete coverage between the lodge, skiing and hiking parties and the base at the airport in Golden, BC. The repeater is powered by a deep cycle battery and a solar charger. The Purcell Lodge system has operated without major problems since start-up. The system provides pollution-free, clean and reliable power in a location where commercial power is not an option. To the visitor, the years of careful planning and the extensive use of high technology are virtually invisible. Without them the rare combination of comfort and wilderness provided at Purcell Lodge would have been impossible.[1]

Editor's Note: This article first appeared in *Home Power* Issue #33 (February/March 1993) and is reprinted with permission of the authors.

4

Powerful Dreams—
Crown Hill Farm's Hydroelectric Plant

JULIETTE AND LUCIEN GUNDERMAN

Several people who have heard about our hydro plant have all had the same questions! Where on earth did you come up with the idea to build a hydroelectric plant on your farm? They also wonder what it takes to design and build a hydro system. Well, it takes an idea first and foremost. Actually, it takes a lot of ideas. It takes frequently waking up in the middle of the night with many ideas and writing them down. It takes water, optimism, more water, diligence, more water, patience, water, practical common sense and the will to succeed.

The idea came 20 years ago while Lucien was attending his high school's ten year reunion at the Von farm in Carlton, Oregon. Dick Von, a logger and farmer, had always wanted to build a hydroelectric system on the farm. The Vons were just getting started with their project at the time of the reunion, so Lucien got to see the beginning phases of the system. Pipelines had been laid, and the powerhouse had been started.

The gears in Lucien's head started turning. Soon after the reunion, and every three or four years, Lucien contacted the local utility

Left: The big lake (6.1 acres) supplies the larger of the two hydroelectric turbines located 170 vertical feet below.

Right: Lucien Gunderman and the twin Canyon Industries turbines

to see what they thought of the idea of a small hydro plant being built and intertied with the utility. Each time the question was asked, the same answer was given: "It's a great idea, but with the low rates that McMinnville Water and Light currently has it's simply not an economical thing to build, and the payback would be too many years to count." McMinnville Water and Light still has one of the lowest electricity rates in the US.

Eighteen years went by. The year was 2000, and an energy crunch was upon us. This energy crisis prompted us to hire an engineer and pursue the hydro project in earnest. The engineer saw promise in the project.

Lucien again contacted the Vons and arranged for a personal visit and tour of their system, which by then had been in operation for nearly two decades. Juliette, Lucien's mother, and Lucien were both hooked after

---

Hydro History — Full Circle

When Crown Hill Farm was started, my parents were the first rural electric customers of the utility on Baker Creek Road. They also supplied cordwood to the utility to operate a steam turbine that was used to power the electric plant for the city of McMinnville in the early years. The plant used a combination hydro/steam turbine. When water was not available in sufficient supply from Baker Creek to turn the Pelton wheel turbine, steam was generated by a wood-fired boiler. This power plant was located just one mile above the farm's entrance.

This plant was a 200 kilowatt system that was the only source of electricity for the city of McMinnville in the early days. It was built in 1907. The vertical head was 237 feet, only 50 feet more than we have on our system. The water was carried via a 24 inch pipeline approximately one mile. A dam was installed, complete with fish ladder and a large vertical slide gate for flushing the dam.

As a child, I played at the dam on countless occasions and was always intrigued with it, the fish ladder and all parts of the system. Recently I got to see the remains of the original turbine that was decommissioned in 1952. When McMinnville Water and Light came on-line with Bonneville Power, it was required that they shut down their own power plant.

Lately there's been talk of recommissioning this plant. The Crown Hill Farm project is a demonstration of the viability of hydroelectricity. Things have come full circle with the completion of Crown Hill Farm's hydro project.

The Baker Creek power plant. Juliette's father delivered wood to it in the early days.

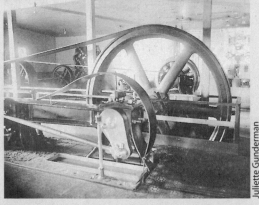

Juliette Gunderman

seeing the Vons' system. We both knew that Crown Hill Farm had the potential for a hydro system if Lucien's ideas could be put together into a finished package.

McMinnville Water and Light

Lucien again made a trip to the local utility, and this time got a much different response. Rates were increasing and were expected to keep increasing, and electricity was now in short supply. Amazing how a few years and an energy crisis can change a situation.

However, no one in the 113 year history of the utility had ever built a grid-tied micro-hydro project in its service area. The company was reluctant to be more accommodating than they were legally required to be. And they were completely unfamiliar with the interface and the induction generation system that we were proposing.

After several meetings with us, the McMinnville Water and Light Commission and the staff of the utility saw the value that local renewable energy would provide to the community. They offered to be more flexible and to install a pole and dual meters for the project. One meter was for incoming electricity and one for generated electricity flowing back into the grid.

Crown Hill Farm Hydro

Our farm was started in 1920 by Damien and Zephirine Mochettaz, Juliette's parents. It encompasses nearly 800 acres. Crown Hill Farm is aptly named — it sits in the rolling hills west of McMinnville, and actually has a crown of high hills around the south and west ends of the property above our main farm buildings and residences. We recently put a conserva-

Greg Wheeler

The two Canyon Industries turbines flank the 30 kilowatt, three phase generator.

tion easement on the farm to protect its natural beauty and open space in perpetuity.

Once we got started on the hydro project, a hydrologist was hired. He determined that approximately 175 acres could be used as the watershed for the project. There were already two lakes on the farm. The large one was built in 1954 and is used for irrigation; it holds roughly 22 acre feet (2,700 cubic meters; just over 7,000,000 gallons) of water. The second lake was quite small, but could act as a collector for the new lake that would be built at a slightly lower elevation. It was in a more beneficial location for collecting and regulating water for this project.

We verified rainfall for the last 50 years, which showed that adequate water would be available (at least in the winter months) to run the system. The farm, located six miles west of McMinnville, gets 46 to 50 inches of rain per year. After doing this research we decided to go ahead with the project, so we started to make arrangements and plans. Various sources were

At the little lake, a blue intake pipe is drilled with hundreds of ¼ inch holes. The galvanized shroud prevents debris from clogging the intake.

Greg Wheeler

used to research the project, from the Internet to Oregon State University.

We dug several small collection ponds and nearly 5,500 feet of ditches. This included ditches to divert water to two reservoir ponds that supply water to the project, as well as for the main lines to the turbines.

Water for this power plant is from upland sources, including several artesian springs that run year-round, and from collected rain runoff. The head (vertical distance the water falls) from the little lake diversion is 85 feet, and the head from the large lake is 170 feet. The water leaving the powerhouse runs into Baker Creek, which borders our property, and eventually into the Yamhill, Willamette and Columbia Rivers.

Intake and Pipe

The penstock system includes two main pipes. The ten-inch pipeline from the big lake runs 1,850 feet and feeds the larger of the two turbines. The eight-inch pipeline from the little lake runs 950 feet and feeds the smaller turbine. Both pipelines are buried five feet deep. Both lines are straight runs except one 45° elbow in each line, where there are thrust blocks — large concrete blocks attached to the pipe. The thrust blocks anchor the pipe and absorb the force of the water on the fittings and pipe. There are also thrust blocks on the ten-inch line where it comes up and into the filter, which is adjacent to the powerhouse.

We did a lot of the manual work and used our backhoe and dozer for much of the excavation. All four small diversion ponds and the second lake were built by Lucien and a friend, Jim Modaffari, who does excavation work. Pipelines were laid by a professional who deals with high pressure irrigation lines all of the time. All pipeline ditches were backfilled by Lucien. We also designed and helped build all portions of the tailrace and powerhouse.

The big lake has a filter screen on the end of a pipe in the lake. The stainless steel screen box is approximately five feet tall by two feet wide by two feet deep and is clamped to the pipe with a steel clamp. It will not allow any particles or debris larger than ¾ inch in diameter to enter the penstock. The penstock is the pipe that delivers the water from the intake to the turbine.

We installed a 2,000 gallons per minute in-line filter that removes any debris that might get through the main screen in the bottom of the lake. This filter has a built-in brush and blow-off valve so that it can be cleaned and flushed even while in operation.

Canyon Industries manufactured our two Pelton turbines. They did not want any particles larger than ¼ inch passing through the turbines. When we ordered the filter, the

---

Technical Specifications

System type: Batteryless grid intertie, three phase, 240 volt, open delta wiring configuration

Static head: Little lake, 85 feet; big lake, 170 feet

Flow rate: 65 to 1,850 gallons per minute

Large turbine: Canyon 1215-2, twin nozzle Pelton, 12 inch pitch diameter

Small turbine: Canyon 9513-2, twin nozzle Pelton, 9.5 inch pitch diameter. Both turbines have manganese bronze runners. Our system is unique in that two turbines actually run one generator. Either turbine can run the generator with one nozzle or two, or both turbines with one, two, three or four nozzles. Valves are automatically opened and closed through the secondary control panel according to lake levels.

We have two on/off setpoints in both lakes according to levels that we can select at the powerhouse. The system can operate completely unattended, with a variety of weather conditions and available flows.

Nozzles: Minimum nozzle size for both turbines is 0.63 inches. Maximum nozzle size for the large turbine is 1.6 inches. Maximum nozzle size for the small turbine is 1.4 inches. Nozzles are fixed-jet type nozzles that are easily changed for seasonal water fluctuations.

Generator: Marathon, M/N 324TTDP7071, 240 VAC, three phase, induction, 60 hertz, 0.5 to 30 kilowatt, 1,800 revolutions per minute, belt driven

Main disconnect fuse/breaker: 100 amps at 240 volts, three phase

System performance metering: PQM, Multilynn. Shows voltage, amperage, instantaneous power output and approximate kilowatt-hours

Average kilowatt-hours per month: 3,000 kilowatt-hours

January-April 2003

Utility residential kilowatt-hour cost: 3.8 cents for the first 1,000 kilowatt-hours; 4.1 cents over 1,000 kilowatt-hours

Percentage offset by system: 50% in first six months of operation

Lucien Gundermann

---

manufacturer opted to use ¹⁄₁₆ inch stainless steel mesh for the screen. They felt that we would have little or no trouble with organic material getting hung up on this size mesh. We've flushed the filter four times since it was installed eight months ago. The filter has two pressure gauges mounted on the body, and it is recommended that it be flushed when there is a five pounds per square inch difference between source and output ports on the filter. When we did flush the filter, it had never reached the five pounds per square inch difference. We just wanted the system to work at

optimum efficiency. The filter has worked very well.

The little lake intake screen is a combination screen and filter. It was made from a piece of schedule 40, eight-inch PVC pipe that is vertical in the lake. Lucien drilled ¼ inch holes for seven hours one day to make this intake filter/screen. It works perfectly. We used some 24 inch galvanized heat duct to make a shield that surrounds the intake pipe. Water must enter at the bottom of the shield, so a very limited amount of debris is pulled up to the actual screen pipe. We plan to build a catwalk this summer so we can run a brush up and down the pipe to dislodge small particles that are sucked against the pipe during operation.

Latest Technology

The system is designed to run with water from one or both lakes in combination and uses two, twin nozzle, Canyon Industries Pelton wheel turbines that are synchronized via a belt drive to operate in unison. The turbines are a fixed-nozzle design, and nozzles can easily be changed depending on the available water. The low-end output of the system is approximately 500 watts, and the high-end rating is 30 kilowatts. The turbines will run efficiently with a volume as low as 65 gallons per minute or as high as 1,850 gallons per minute.

Since we have a net metering contract with

McMinnville Water and Light, we cannot exceed 25 kilowatts at any time during the year. Oregon law for net metering limits the output to 25 kilowatts. This system is a three phase, 240 volt grid interface system that is wired directly into McMinnville Water and Light transmission lines. Electricity generated is credited by the utility through a net metering agreement signed earlier in the year.

This hydro plant incorporates the latest technology and has automated features that monitor lake levels and temperature, as well as generator function, frequency, kilowatt output and power quality. It has automated controls that open and close valves on the turbines according to lake levels. Level options are programmed on a keypad on a secondary control panel, and a digital readout gives water levels in both lakes as well as a temperature reading of the lakes and Baker Creek.

The system also has a dual timer option for turbine operation and a manual override for select situations. It has six fail-safe controls that will automatically shut down the generation system when necessary. These controls protect the equipment and ensure that no elec-

Left: The main contactor enclosure contains the overcurrent protection, manual disconnect and main relay.

Right: Two kilowatt-hour meters measure grid energy to the farm and hydro energy to the utility grid. Below is the disconnect required by McMinnville Water and Light. The shed that houses the turbines and controls is in the background.

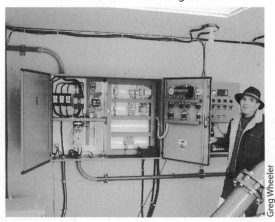

Greg Wheeler

Greg Wheeler

tricity will flow into the local utility lines when they are being repaired by utility workers during a grid outage. An automatic water shut-off feature will turn off all water to the turbines in the event of a utility failure or any other system malfunction. Lucien designed this feature so that a valuable resource — water — will not be wasted.

Taking Care of the Fish

To protect the fish the system monitors the temperature of the water and the levels of each

---

Induction Generation

When we looked at what type of generating system to use, the most economical and easily grid-intertied system was an induction generation system. With an induction system, you use a regular induction motor. When the turbine spins the motor shaft, the motor becomes a generator and generates instead of consuming electricity.

Synchronization

One beauty of this system is the simple controls required to connect to the grid. There is an electronic tachometer that monitors the system speed. When the induction motor/generator hits generating speed, the control panel connects it to the grid. After the phases line up, the grid locks the generator at 60 hertz, and it is generating. Induction generators are easily obtained, although ours, because of the double turbine drive, had to be special ordered.

Safety

Within the intertie panel are relays that will sense a grid failure and automatically shut down our system. So the system cannot endanger utility crews with an unanticipated electrical feed. Electrical codes are very specific as to auto shutdown in the event of a grid failure. The utility was very concerned about the intertie, since we are the first non-utility hydro plant to come on-line in their history. If the utility needs to work on main transmission lines, they can shut off our system with a manual disconnect switch, which in turn shuts down the generating system in a matter of seconds.

Induction or Asynchronous?

We did also consider using an asynchronous generator. With an asynchronous system, we would have had the option of complete stand-alone power, even if the grid went down or failed. But it also would have meant more wiring, a grid-interactive inverter, batteries and a high voltage utility disconnect. The additional cost would have been approximately US$20,000 to $25,000. We decided that since we already had backup generators, this additional investment really made little sense.

Transmission

Our transmission lines to tie in with the grid are only about 450 feet of overhead run, and we were able to tie into our existing 240 volt, three phase, open delta irrigation service. This kept the cost down and is working very well.

lake. If the levels get too low or the water is too warm, it can't be run through the turbines and introduced into Baker Creek. Each lake has a two-stage setting with level controls to maximize resource usage and allow for automatic control of generator output.

Water leaving the turbines through the concrete tailrace is slowed to alleviate erosion and eliminate water turbulence when it merges with Baker Creek. The water is aerated thru a series of diversion bars of expanded metal, oxygenating the water to facilitate fish habitat in Baker Creek. The creek is listed as a fish-bearing stream, which includes such species as dace, sculpin, cutthroat trout, lamprey, crayfish, winter steelhead and coho salmon.

The temperature standard for cutthroat trout, steelhead and salmon is quite cold. Technically, the temperature is not to exceed 55°F May 1 through July 15, 65°F July 16 through October 15 and 55°F October 16 through 31. We did a lot of talking about these

Water Wheels

A small, four-foot water wheel is incorporated into the design of this hydro project, just for fun. The water wheel was constructed from an antique steel wheel that was 36 inches across, and the ½ gallon buckets were from a dismantled feed mill, the old Albers Mill on Front Street in Portland.

For the time being, it is just a functioning, aesthetic addition to the project. The dry seasons during these last two years have also limited the possibilities for the water wheel. Water for the water wheel comes from a third source of water so it does not take water away from the main turbines.

A new, small water wheel, made from an old squirrel cage fan, mounted on a frame on roller bearings has been constructed and direct coupled to a permanent magnet motor from a computer drive. This small water wheel is mounted on top of the original tailrace, and is supplying a continuous output of between ½ to 1½ amperes at 12 VDC to two automobile batteries. A 700 watt inverter supplies emergency lighting, runs a battery charger for the DC portion of the main project and runs some decorative lighting on the exterior of the hydro building.

This portion of the project was just a brainstorm and a fun part of the overall project. It uses the same water that is fed to the larger water wheel, so it does not take away from the main turbines. The inspiration for this project was a similar picohydro system at otherpower .com/otherpower_experiments_waterwheel .html.

A small, four-foot diameter water wheel was built using parts from an old feed mill.

Greg Wheeler

Edmonton Public Library
Mill Woods
Express Check #4

Customer ID: **********6890

Items that you checked out

Title: Serious microhydro : water power
solutions from the experts
ID: 31221096084947
Due: August 21-18

Total items: 1
Account balance: $0.00
July-31-18
Checked out: 3
Overdue: 0
Hold requests: 0
Ready for pickup: 0

Thank you for visiting the Edmonton
Public Library

www.epl.ca

temperature criteria with state and federal agencies and ended up having to install the temperature monitoring equipment and setting it up for auto shutdown if we exceed the creek temperature.

Alarms and Controls

Lucien installed signal lighting that is visible from our home to show when the system is operational. An audible and visual alarm were also designed and installed to alert us of a grid failure or system shutdown. A low voltage power supply and low voltage actuators are used for the auto valve control that is run from the head level sensors in each lake. The actuators were purchased from Burden's Surplus Center in Minnesota. A 24 VDC battery charger charges two small 12 volt batteries in series, which allows the system to close valves in the event of a grid failure. If the grid goes down, a relay simply tells the panel to close all valves and shut off the water.

Even though Lucien thought up many of the details in this system, the turbines, panels and most of the control mechanisms were not manufactured on-site. There were countless phone calls and e-mail messages, as well as continuing research into many of the details of this project. It would not have come together without the help of Canyon Industries, Bat Electric and Inertia Controls. The equipment used for the project has a long life and is expected to perform for a century or more with little maintenance.

Determination

This 30 kilowatt capacity microhydro plant is the first newly licensed hydro plant in the state of Oregon in the last 20 years. Several people

Hydro System Costs	
Turbines and switchgear	$40,000
Pipe and installation	$22,500
Building and concrete	$10,000
Excavation	$7,500
Water wheels (both) and equipment	$500
Electrician	$6,500
Engineering and miscellaneous costs	$2,500
Hydrologist report	$1,600
Total	**$106,600**
Note: all figures in US dollars	

Lucien Gunderman

have told us that they would not have had the determination, persistence and patience to deal with all of the agencies and their rules, regulations, restrictions and timetables.

Local, state and federal agencies that had jurisdiction or commented on the project included: Yamhill County Planning, Oregon Department of Fish and Game, Oregon Department of Environmental Quality, Oregon Division of State Lands, Oregon Department of Forestry, Oregon Department of Agriculture, Oregon Office of Energy, State Historic Preservation Office, Oregon Department of Water Resources (the lead agency), Northwest Power Planning Council, Oregon Parks and Recreation Department, Oregon Land Conservation and Development Department, Oregon Public Utilities Commission, US Department of Fish and Wildlife and the National Marine Fisheries Service. Our commitment was tested when dealing with these agencies during the licensing process, which took a year and a half.

Benefits

We see many benefits from our hydro project. The major one is supplying electrical energy.

Crown Hill Farm's Dual Turbine Hydro System

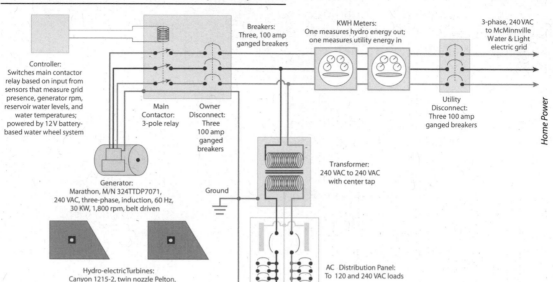

Controller:
Switches main contactor relay based on input from sensors that measure grid presence, generator rpm, reservoir water levels, and water temperatures; powered by 12V battery-based water wheel system

Breakers:
Three, 100 amp ganged breakers

KWH Meters:
One measures hydro energy out; one measures utility energy in

3-phase, 240 VAC to McMinnville Water & Light electric grid

Main Contactor:
3-pole relay

Owner Disconnect:
Three 100 amp ganged breakers

Utility Disconnect:
Three 100 amp ganged breakers

Generator:
Marathon, M/N 324TTDP7071, 240 VAC, three-phase, induction, 60 Hz, 30 KW, 1,800 rpm, belt driven

Ground

Transformer:
240 VAC to 240 VAC with center tap

Hydro-electric Turbines:
Canyon 1215-2, twin nozzle Pelton, 12 inch (30.5 cm) pitch diameter; Canyon 9513-2, twin nozzle Pelton, 9.5 inch (24 cm) pitch diameter

AC Distribution Panel:
To 120 and 240 VAC loads

Home Power

The system is expected to generate 96,000 kilowatt-hours of electricity per year, enough to supply approximately eight typical Oregon homes. It should generate enough electricity to meet all of Crown Hill Farm's electricity — and about 25% more, which McMinnville Water and Light will buy at wholesale and resell to other consumers.

Another important benefit is that our project is a renewable resource and does not deplete any natural resources. The two main lakes already existed and needed no structural changes. The diversion ponds provide additional wildlife habitat. The project adds cold, aerated water to Baker Creek, which enhances fish habitat.

"A hidden benefit is that this little hydro system is actually improving the power quality for their neighbors," said Christopher Dy-

mond of the Oregon Office of Energy. "Lucien and Juliette's investment in local clean energy reflects both their patriotism and good stewardship." One of the main pipelines also incorporates irrigation risers that will add efficiency to summer irrigation because of the larger supply line with more pressure. The large lake has been used for irrigation purposes since 1954.

The project better controls runoff water. It collects and diverts water to the new diversion ponds and two regulation lakes. This dramatically reduces erosion, sedimentation and water damage to drainage ditches and Baker Creek.

Hydro Dreams

It has been said that dreams come and go. In this case, our dream has come true, especially for Lucien, who never gave up hope on the idea that our farm and its natural resources could one day be used to supply electricity to ourselves and others.

A project like this is a big undertaking, with many unexpected costs and hurdles along the way. But the feeling of satisfaction, pride and good stewardship is well worth the time, energy and hard work to bring it all together. It is a great feeling to see a project come together and work after dreaming and planning for many years.[1]

Editor's Note: This article first appeared in *Home Power* Issue #96 (August/September 2003) and is reprinted with permission of the authors.

PART II

Household Pressure Sites

These many sites run on far less pressure than the classic systems. Gravity water systems are often used, when available, to deliver domestic water to houses. Many find that the same pipe that delivers water to the house has enough capacity left over to generate power. Seeing this potential and creating the technology to realize it was the particular genius of Don Harris in the US and Paul Cunningham in Canada.

Following is a selection of sites that run on very approximately as much pressure as your kitchen sink. For example, our current urban household pressure is often at 90 pounds per square inch (psi), the equivalent of well over 200 feet of head.

Some useful examples of this kind of system didn't make it into this collection in their full form, but they are summarized in Figure 1 (Case Studies by Head) in the Introduction. In *Home Power* Issue #71, there is an excellent article on the use of induction motors for small hydro, as well as several case studies, in the article by the late Bill Haveland titled "Induction Motors for Small Scale Hydro." There's a humourously written case study called "Off-Grid Pioneers," reassuring the reader that few indeed are the sacrifices made when powering your remote homestead with microhydro, in *Home Power* Issue #59. In addition to these, a couple of microhydro systems are mentioned in "Renewable Energy: Kiwi Style" in Issue #49.

5

The Small AC System

SCOTT L. DAVIS

Part of this article appears as a case study in the RETScreen Clean Energy Project Analysis Software.[1] This chapter expands upon the lessons learned at this site.

Results

A microhydro system was built for an off-grid family ranch in a remote area west of Lillooet, BC over a period of time in the 1990s. It was put together with local talent and quite a bit of volunteer labour from all concerned. It replaced an existing homemade microhydro system, taking advantage of an existing gravity water system, with a more efficient and reliable system offering more power. This new system provided electrical service and also furnished domestic hot water and a significant amount of space heating.

System Description

The system is based on a Canadian turbine, the Energy Systems and Design Turgo with four inch runner, well-tested in battery charging applications where it generates up to 1.5 kilowatts. It is perfectly capable of generating more than three kilowatts, as it does in this three-jet system. It drives a 12 kilowatt brushless AC alternator. Although larger than strictly necessary, the alternator proved to be reasonably efficient, and the extra mass came in handy acting as a flywheel to help in motor starting.

Lessons Learned
- Every site is unique.
- The supply of renewable energy from the

This runner from a battery charging turbine is robust enough to generate a few kilowatts.

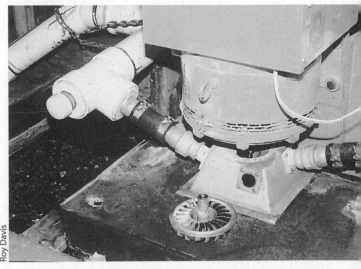

Roy Davis

microhydro system greatly exceeds the home's demand.

- This system performed slightly better, in terms of satisfying sensitive and transient loads, than a battery/inverter system rated at 2.5 kilowatts — and also provided lots of heat.
- This direct AC system does not require a careful assessment of the demand, as might a microhydro battery charging system, and is simpler and more reliable than systems incorporating batteries.
- Maintenance requirements on this system are minimal. Bearings need to be replaced every seven years. The intake needs to be cleaned, a 15 minute task monthly or, during fall, even weekly.

The Big Picture

Microhydro technology can be by far the most cost-effective solution to the problem of providing basic electrical service to an off-grid residence, even when the system output seems very small indeed. Europeans and many off-grid consumers can testify that the first 100 kilowatt-hours per month makes possible the majority of benefits associated with electric service, and that a couple of 100 kilowatt-hours per month permits a higher standard of living. These same consumers often report that a 2.5 kilowatt inverter adequately meets their household power requirements. Although the homeowner could have invested more money, effort and engineering to generate more power from the available water resource, a 3.2 kilowatt system was most satisfactory. Despite the high heating requirements of the log home, excess power was being dumped for much of the year. Lack of opportunities for getting an

A third jet gave more space heating in the winter.

economic return for excess power often limits the size of projects that can be justified. An opportunity to create value from surplus power would be welcomed by many microhydro users.[2]

The History of This Small AC System

This site has seen many upgrades over the years. Each one taught many lessons about just what was, and was not, possible with various small AC systems. At the time, and even today, you will often hear that an AC system has to be at least a couple of kilowatts in size to be practical. However, when we bought it, this ranch came with a 600 watt AC system that ran a perfectly ordinary refrigerator as well as lights. We thought it was the best system around. The output was a true sine wave, running electronics and other demanding small loads quite well. This held, even though the frequency dropped to 50 cycles or lower as the system became overloaded. The refrigerator ran for years and years like this. Incandescent lightbulbs (this was before compact fluorescent bulbs were available) seemed to last much longer at lowered voltage.

It's also possible to hear that control of AC systems is difficult without expensive controllers. The original system was entirely uncontrolled, without having to leave all the lights on or to move quickly when loads changed. The person who put this system together was quite a clever guy mechanically, really the classic rancher. At a scrapyard he found a big old turbine that had been in use for decades. It was so big that even with over 100 feet of head, it only turned a few hundred revolutions per minute. The pipe had many kilowatts of potential, and so the rancher was clever enough to trade output for controllability. Instead of gearing the wheel to the alternator to get the most power with a large diameter pulley, he used a smaller pulley than required. This didn't slow the turbine down enough to make much power, but on the other hand, it didn't run away to high revolutions per minute when the loads were off either.

When our family bought the ranch, I noticed that there was little difference between the static head and the pressure when the turbine was running. We increased the nozzle size and used more water. I replaced the pulley with a flywheel from a hay baler, both for larger diameter and for the flywheel effect.

The system was still operating far below its potential, and so controlling the system was just not that difficult. The mass of the fly-

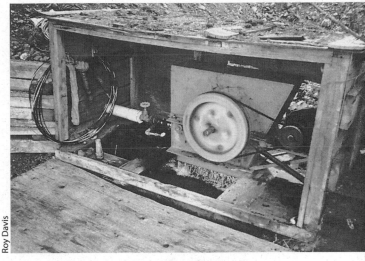

A scrapyard system with many clever features

wheel, plus the fact that the gearing was still not ideal, meant that speeds changed slowly. As the system got loaded up, the wheel slowed down more and became more efficient. Then, of course, it got less efficient as it speeded up, the cups kind of running away from the water, and so the system tended to be somewhat self-correcting. The controller now consisted of a voltage sensitive relay running a water heater. When the voltage got too high, the inexpensive relay cut in the water heater. When the voltage was too low, it cut off the load. We used a dimmer to adjust the apparent size of the load so that the whole thing ran really quite smoothly.

This upgrade made a system of about 1,000 watts. In addition to the fridge, we now had a freezer and automatic washer with soft start kits.

The point is that we had many modern conveniences with a 600 watt system, to be sure. None of the appliances were anything but ordinary until we got the soft start kit installed on the new freezer and washer. Even at 600 watts, we could easily have run a Sunfrost high efficiency fridge and freezer, and found or made a custom washer. It's just that in this case, which I now see as quite unusual, money was always better spent making more of the potential of the pipe. We didn't even have solar hot water, since we could get quite a bit of hot water even with 1,000 watts.

After all, a 1,000 watt system delivers nearly as many kilowatt-hours per month as the average local household uses. There were problems with starting larger tools, but this again wasn't really a hardship.

Reliability drove upgrades as much as the desire for more power. One morning, the lights seemed kind of dim. When we went to the powerhouse, one of the bearings had failed, spilling bearings on the floor. The other bearing, however, still held up. Ball bearings all over the floor, and it still put out power!

At first, only two jets were used in the new turbine. It still gave 2,500 watts, two and a half times as much as before. The hot water heater worked a lot better, we could start anything that an inverter rated at 2,500 watts could start and maybe a little more, and we got quite a bit of space heating in the winter.

We found that the first small system actually delivered quite a high standard of living. The upgrade to an output of 1,000 watts allowed us to continue to use inexpensive appliances and forego solar water heating. The upgrade to a new direct drive 2,500 watt system gave blessed reliability from its one moving part, as well as much more heat. The 3,200 watt upgrade provided even more space heating during the winter months.

There was still more potential, up to five kilowatts, in the existing pipe. These last kilowatts would require spending real money for a more efficient turbine, fixing up the intake for higher flows and other improvements. We had finally reached a point of diminished returns, and just enjoyed it.

6

Kennedy Creek
Hydroelectric Systems

RICHARD PEREZ

In the 6,000 foot Marble Mountains of Northern California, it rains. Wet air flows straight from the Pacific Ocean only 40 airline miles away. This moist ocean air collides with the tall mountains and produces over 60 inches of rainfall annually. Add this rainfall with the spectacular vertical terrain and you have the perfect setting for hydroelectric power. This is the story of just one creek in hydro country and of five different hydro systems sharing the same waters.

Home Power

Kennedy Creek

Kennedy Creek is on the west drainage of 4,800 foot Ten Bear Mountain. The headwaters of Kennedy Creek are located in a marsh at 2,500 feet of elevation. The headwaters are spread out over a ten acre area, and the power of Kennedy Creek doesn't become apparent until its waters leave the marsh. After a winding course over five miles in length, Kennedy Creek fi-

nally empties its water into the Klamath River at about 500 feet elevation. This gives Kennedy Creek a total head of 2,000 vertical feet over its five mile run.

The volume of water in Kennedy Creek is not very great. While we weren't able to get really hard data as to the amount of water, the residents guessed about 500 gallons per minute. Kennedy Creek is not large by any standards. It varies from two to eight feet wide and from several inches to about four feet deep. We were able to cross it everywhere and not get our feet wet. The point here is that you don't need all that much water if you have plenty of vertical fall.

The Kennedy Creek Hydro Systems

Kennedy Creek supports five small scale hydroelectric systems. Each system supplies electric power for a single household. Each system uses the water and returns it to the creek for use by the next family downstream.

These systems are not newcomers to the neighborhood; they have been in operation for an average of 7.6 years. These systems produce from 2.3 to 52 kilowatt-hours of electric power daily. Average power production is 22 kilowatt-hours daily at an average installed cost of US$4,369. If all the hydroelectric power produced by all five Kennedy Creek systems is totaled since they were installed, then they have produced over 305 megawatt-hours of power. And if all the costs involved for all five systems are totaled, then the total cost for all five systems is US$21,845. This amounts to an average of seven cents per kilowatt-hour. And that's cheaper than the local utility. One system, Gene Strouss', makes power for three cents a kilowatt-hour, less than half what's charged by the local utility.

All the power production data about the Kennedy Creek hydroelectric systems is summarized in the table. All cost data is what the owners actually spent on their systems. Being country folks, they are adept at shopping around and using recycled materials. The cost figures do not include the hundreds of hours of labor that these hydromaniacs have put into their systems.

Let's take a tour of the Kennedy Creek hydros starting at the top of the creek and following its waters downward to the Klamath River.

Gary Strouss

Gary Strouss wasn't home the day that Bob-O, Stan Strouss and I visited Gary's hydroelectric site. Gary is a contractor and was off about his business. So as a result, we got this info from his brother Stan and father, Gene (the next two systems down Kennedy Creek).

Gary's hydroelectric system uses 5,300 feet of four-inch diameter PVC pipe to deliver Kennedy Creek's water to his turbines.

Hydroelectric System Owner	System's Age in Years	Average Power Output (watts)	Daily Power Output (kilowatt-hours)	Total Power Made (kilowatt-hours	System Cost	System Power Costs to Date ($ per kilowatt-hour
Gary Strouss	6	2,040	49	107,222	$8,795	$0.08
Stan Strouss	8	180	4	12,614	$3,520	$0.28
Gene Strouss	9	2,166	52	170,767	$5,950	$0.03
Max and Nena Creasy	6	97	2	5,096	$1,295	$0.25
Jody and Liz Pullen	9	120	3	9,461	$2,285	$0.24
Averages	7.6	921	22	61,033	$4,369	$0.18
Totals	38	4,603	110	305,163	$21,845	
Total average cost per kilowatt-hour from Kennedy Creek systems						**$0.07**

Richard Perez

Note: All figures in US dollars

The head in Gary's system is 280 feet. Static pressure is 125 pounds per square inch at the turbines.

Gary uses two different hydroelectric generators. One makes 120 VAC at 60 hertz directly and the other produces 12 VDC. The 120 VAC system is very similar to the one his father Gene Strouss uses and is described in detail below. Gary's 120 VAC system produces 3,000 watts about eight months of the year. During the summer dry periods, Gary switches to the smaller 12 volt hydro.

The 12 volt DC system uses a Harris turbine that makes about 10 amperes of current. The Harris turbine is fed from the same pipe system as the larger 120 VAC hydro.

Gary's home contains all the electrical conveniences, including a rarity in an AE powered home — an air conditioner! The 120 VAC hydro produces about 48 kilowatt-hours daily, so Gary has enough power for electric hot water and space heating.

Stan Strouss

Stan's hydro is supplied by 1,200 feet of two-inch diameter PVC pipe. His system has 180 feet of head. In Stan Strouss' systems, this head translates into 80 pounds per square inch of static pressure and 74 pounds per square inch of dynamic pressure into a 7/16 inch diameter nozzle.

Stan uses a 24 volt DC Harris hydroelectric system producing three to ten amperes. Stan's hydro produces an average of 180 watts of power. This amounts to 5,400 watt-hours daily. The system uses no voltage regulation. The DC power produced by the hydro is stored in a 400 ampere-hour (at 24 VDC) C&D lead-acid battery. These ancient cells were purchased as phone company pull-outs eight years ago. Stan plans to use an inverter to run his entire house on 120 VAC. Currently, he uses 24 VDC for incandescent lighting. When I visited, there was a dead SCR type inverter mounted on the wall, and Stan was awaiting delivery of his new Trace 2524.

Stan's system is now eight years old. The only maintenance he reports is replacing the brushes and bearing in his alternator every 18 months. That and fixing his water intake filters wrecked by bears. Stan and his father, Gene, own and operate a sawmill and lumber business from their homesteads. This business, along with raising much of their own food, gives the Strouss families self-sufficiency.

Gene Strouss

Gene Strouss' hydroelectric system is sourced by 600 feet of six-inch diameter steel pipe connected to 1,000 feet of four-inch diameter PVC pipe. Gene got an incredible deal on the 20 foot lengths of steel pipe, only US$5 a length.

A 12 inch diameter horizontal cast steel Pelton wheel translates the kinetic energy of moving water into mechanical energy. The Pelton wheel is belted up from one to three and drives an 1,800 revolutions per minute, 120 VAC, 60 hertz AC alternator. All power is produced as 60 cycle sinusoidal 120 VAC. The Pelton's mainshaft runs at a rotational speed of between 600 and 800 revolutions per minute. The output of the alternator is between 1,500 to 2,500 watts depending on nozzle diameter. At an annual average wattage of 2,000 watts, Gene's turbine produces 48,000 watt-hours daily.

The pipe delivers 60 pounds per square inch dynamic pressure into a 9/16 inch diameter nozzle, for summertime production of 1,500 watts at 70 gallons per minute of water through the turbine. In wintertime with higher water levels in Kennedy Creek, Gene switches the turbine to a larger, $^{13}\!/_{16}$ inch diameter nozzle. Using the larger nozzle reduces the dynamic pressure of the system to 56 pounds per square inch and produces 2,500 watts while consuming 90 gallons per minute.

Gene's system is nine years old. The only maintenance is bearing replacement in the alternator every two years. Gene's system uses no batteries; all power is consumed directly from the hydro. Gene keeps a spare alternator ready, so downtime is minimal when it is time to rebuild the alternator. Regulation is via a custom-made 120 VAC shunt type regulator using a single lightbulb and many parallel connected resistors. Major system appliances are a large deep freezer, a washing machine, 120 VAC incandescent lighting and a television set.

Gene's homestead is just about self-sufficient (which is why he needs his freezer). Hundreds of Pitt River rainbow trout flourish in a large pond created by the Pelton wheel's tailwater. The trout love the highly aerated tailwater from the hydro turbine. Gene grew 100 pounds of red beans for this winter and maintains two large greenhouses for winter vegetables. Gene Strouss also keeps a large apple orchard. Gene raises chickens and this, with the trout, make up the major protein portion of his diet. His major problem this year was bears raiding the apple orchard and destroying about half of the 250 trees. For a second course, the bears then ate up over 60 chickens, several turkeys and a hive of honey bees. Gene called his homestead, "My food for wildlife project."

Max and Nena Creasy

Seven hundred feet of two-inch diameter PVC pipe sources a Harris hydro turbine with two input nozzles. Static pressure at the turbine is about 80 pounds per square inch from a vertical head of 175 feet. It produces five to eight amperes depending on the availability of water.

Max and Nena use 100 feet of #2 USE aluminum cable to feed the hydro power to the batteries. Max and Nena's system uses two Trojan L-16 lead-acid batteries for 350 ampere-hours of storage at 12 VDC. All usage is 12 volts directly from the battery. Max and Nena don't use an inverter. The system uses no voltage regulation, and overcharging the batteries has been a problem. Power production is 97 watts or 2,328 watt-hours daily.

The major appliances used in this system are halogen 12 VDC incandescent lighting, television, tape deck and amplifier. This system has been operation for the last six years. Nena reports two year intervals between bearing and brush replacement in their alternator.

Max works with the US Forest Service, and Nena runs a cottage industry making and selling the finest chocolate truffles I have ever eaten.

Jody and Liz Pullen

Jody and Liz's hydro system uses 1,200 feet of two-inch diameter PVC pipe to bring the water to the turbine. Jody wasn't sure of the exact head in the system, and without a pressure gauge it was impossible to estimate. The system works, producing more power than Jody and Liz need, so they have never investigated the details.

The turbine is a Harris 12 volt unit. Jody normally sets the Harris current output at six to ten amps so as not to overcharge his batteries. An average output figure for this system is about 120 watts or 2,800 watt-hours daily. The power is carried from the hydro to the batteries by 480 feet of #00 aluminum USE cable.

The batteries are located in an insulated box on the back porch. The pack is made up of four Trojan T220 lead-acid, golf cart batteries. The pack is wired for 440 ampere-hours at 12 VDC. This system uses no voltage regulation, and Jody has to be careful not to overcharge the batteries. Jody uses all power from the system via his Heart 1000 inverter. He also uses a gas generator for power tools and the washing machine. These tools require 120 VAC and more power than the 1,000 watt inverter can deliver.

Jody and Liz have used this hydro system for their power for the last nine years. They report the same biannual alternator rebuild period. Jody runs a fishing and rafting guide business on the Klamath River called Klamath River Outfitters.[1] Liz is just about finished her schooling and will soon be a Registered Nurse.

What the Kennedy Creek Hydros have Discovered

Hydroelectric systems are more efficient the larger they get. The smaller systems have higher power costs. The largest system, Gene Strouss', operates at an incredibly low cost of three cents per kilowatt-hour. And that's the cost computed to date. Gene fully expects his hydro system to produce electricity for years to come.

Maintenance in these systems is low after their initial installation. While installing the pipe takes both time and money, after it's done it is truly done. The only regular maintenance reported was bearing and brush replacement and trash rack cleaning. The battery-based DC hydros all showed signs of battery overcharging. Voltage regulation is the key to battery longevity in low voltage hydro systems.

A Parting Shot

As Bob-O and I were driving down Ti Bar Road on our way home, we passed the Ti Bar Ranger Station run by the US Forest Service. They were running a noisy 12 kilowatt diesel generator to provide power for the ranger station. Which is strange because they are at the very bottom of the hill with over 2,000 feet of running water above them. And they have five neighbors above them who all use the hydro power offered by the local creek.

The practical and effective use of renewable energy is not a matter of technology. It is not a matter of time. It is not a matter of money. Using renewable energy is just doing it. Just like the folks on Kennedy Creek do.[2]

Editor's Note: This article first appeared in *Home Power* Issue #20 (December 1990–January 1991) and is reprinted with permission of the author.

Independent Power
and Light!

DAVID PALUMBO

When we decided to make our home in the beautiful Green Mountains of northern Vermont, we had no idea where this new adventure would take us. Looking back at our decision of six years ago to produce our own electricity for our new homesite, I am amazed at how this one choice had such a profound effect on our lives.

The Palumbo Family

Our family is comprised of my wife Mary Val, our son Forrest (four years old), our daughter Kiah (two years), our latest addition Coretta (ten months) and myself.

Mary Val and I purchased land in Hyde Park, Vermont during the summer of 1984. At that time we began researching the alternatives to paying the local utility US$6,000 to connect us to their line half a mile away. We were encouraged by friends who produced their own power and a visit to Peter Talmage's home in Kennebunkport, Maine. We decided to "take the road less traveled, and that has made all the difference" as Robert Frost (a Vermonter) put it so well. Talmage Engineering supplied the majority of the hardware, and

Peter answered my questions. We now use alternative energy at all three of our buildings. Let's look at each in turn, as they occurred in time.

The Cherry House System

In the spring of 1985, while living out of a tent, we built what we call the Cherry House. This is first of three buildings designed by M. B. Cushman Design of Stowe, Vermont. The Cherry House is a two-storey saltbox with 950 square feet of living space, heated by a small wood stove. Power for constructing the Cherry House was supplied by a Winco 4,000 watt, slow speed, engine/generator that runs on propane. Energy consumption for the completed house was estimated at 1,300 watt-hours per day. As our primary power source we purchased ten Solenergy 30 watt PV panels that were on the market as seconds in early 1985. The array was cost-efficient, but not really large enough to satisfy our growing power needs. Our battery bank, for the Cherry House, consists of eight Surette T-12-140 deep cycle lead-acid batteries totalling 1,120 ampere-hours at 12 volts. Our loads for this house included our

Dometic 12 VDC refrigerator/freezer (seven cubic feet). We added rigid insulation to reduce the Dometic's power consumption to 420 watt-hours per day. Other loads in the Cherry House include a variety of REC Thin Lite DC fluorescents and a ten-inch Zenith color TV set consuming 4.5 amperes at 12 volts.

When our children began arriving, we added a washing machine and a clothes dryer. The washer and dryer are powered by the Winco generator through the automatic transfer switch built into our Trace 1512 inverter/charger. The transfer switch and charger in the Trace inverter allow us to charge our battery bank and wash the diapers at the same time, all powered by the Winco generator.

The Trace 1512 could not handle the surges of the washing machine. The newer model Trace 2012 will handle most washing machines. We used the Winco propane-fired generator to do the laundry and to help our undersized PV array charge our batteries. The generator was also essential (until we later developed our microhydro site) because we are located in one of the cloudier parts of the country. For example, during our first November here we had one day of full sun followed by a delightful December with three full days of sunshine. Wow! We eventually decided to add a hydro system, since rainfall is generally plentiful here, and our site has the elevation differential to support the hydro.

The Barn and Shop System

During the summer of 1987 we built The Barn with three horse stalls, a 500 square foot workshop and plenty of storage space on the second floor. The Barn is located 450 feet from the Cherry House and 250 feet from the site for the Big House. The distances between these buildings presented us with two choices for the overall power plan. First, we could centralize a battery bank and inverter large enough to handle all of our power needs via 115 VAC. The second choice was to have a separate battery bank in each of the three buildings. I went with the second option because we were building incrementally, and the *whole* was only a fuzzy image in my mind's eye early in the project. Also, I was entering a new business as a designer and installer of alternative power systems. The added experience of three separate systems was desirable and influenced my decision.

Three separate systems may not be the most efficient way to go. I am presently working on another large remote site, with three buildings, several miles north of our land. This installation will take advantage of the products available today. Specifically, NiCad batteries and a powerful inverter located in the garage/shop will serve as the power center for all three buildings. The advantages of this approach include saving time and money in wiring and the ability to use a higher battery bank voltage. This higher system voltage allows the charge source (in this case, PVs) to be located further from the batteries without using the more costly, large diameter wires. For this site, I am designing the system with a 48 volt battery bank.

Our Barn's power system consists of four Trojan L-16W deep cycle, lead-acid batteries with a capacity of 700 ampere-hours at 12 volts. We are using the Heliotrope PSTT 2,300 watt inverter. This inverter has worked well in the shop, powering all of the tools expect those requiring 240 VAC, which are sourced by the generator. We sold the 4,000 watt Winco and replaced it with a Winco 12,500 watt, slow-speed, propane engine/generator. We did this because our carpenter needed to use a high powered air compressor with a six-horsepower electric motor. The other machines powered by the generator include a large table saw, an eight-inch planer and a six-inch joiner. We wired the big Winco so that we are able to turn it on or off from any of the three buildings, using remote four way switches activating the 12 volt solenoid and starter switch at the generator. The remainder of the electrical loads in the Barn/Shop are all lighting. We used Thin-Lite brand DC fluorescents throughout and are very happy with them. Since the shop is the only heated space in the Barn, cold weather light operation was a must. The Thin-Lites work well in the cold. They are efficient: for example they produce 3,150 lumens from a standard 40 watt fluorescent tube. At 78.7 lumens per watt, this is 25% higher than the highly praised PL lights. The 40 watt tubes are inexpensive, locally available and come in a wide variety of spectral outputs.

The Heliotrope inverters do not contain battery chargers (like the Trace models). We use a Silver Beauty battery charger that charges the Trojans quite well from the Winco. However, this battery charger must be turned on with a timer switch as it doesn't have the programmable features of the sophisticated charger built into the Trace inverter/chargers.

The Big House

We felt a traditional, New England, colonial home design offered the features we wanted at a reasonable cost. We were looking for a lot of space, energy efficiency and country charm.

By using all of the space under the roof, we have been able to build a home with 5,300 square feet of heated space. All of this sits on a footprint of 2,160 square feet. The Big House has a full basement (except under the garage) that houses the boiler room, the battery and control room, a large play area for the kids and the cold, root and wine cellars. Without cramping, we can store up to six cords of wood in the basement to augment the woodsheds outside the garage, which hold seven cords.

We have over 100 acres of good forest land that we are managing for both timber production and wildlife habitat. Our woodlots have a sustained yield of over one cord per acre per year to supply our buildings with heat from this renewable resource. Big House's heat is produced by an Essex Multifuel boiler rated at 140,000 Btu. We use it as an oil burner only very occasionally; it is mostly fuelled by wood. The Essex has a ten cubic foot firebox and cycles on and off to satisfy thermostats in the four heating zones within the Big House. The Essex burns by a gasification process and is 95% efficient on wood while producing no creosote emissions. We also get all of our domestic hot water from this 1,500 pound beast's two six gallons per minute heating coils. We use about 15 cords of hardwood per year to heat the Big House and its water. I hope to install a solar hot water heater soon so I can take a summer vacation from loading firewood and shovelling ashes out of the Essex.

Microhydro

I began to think about water power after the first rainy fall of 1985, and by 1987 we began work on our microhydro project. We built a pond on the highest site on our property. The pond is situated on ideal soils (heavy silt on top of glacial hardpan) for pond construction. Our pond is kept full by below-surface springs and surface runoff.

The pond's surface is 210 feet in elevation above our turbine. The pipeline (penstock) is buried under the pond's dam and to a depth of four feet for its entire 1,250 foot run. The inside diameter of the pipe is two inches. I vary the water's flow rate depending on how much power we need, while trying to keep the pond reasonably full. By changing the hydro's nozzle from ¼ to ⅜ inches, I change the flow rate from 17.5 to 38 gallons per minute.

The turbine is an Energy Systems and Design IAT-1½ Horsepower Induction Generator. It was chosen for this application because of the cost of the long wire runs going from the turbine building to the three buildings. The induction generator makes three phase, high voltage AC current, and the higher voltage requires smaller gauge wire on long runs. The longest of these runs is 450 feet to the Cherry House. In retrospect, I would have been better off swallowing the additional expense of larger wires (US$600) and going with a 24 volt DC high output alternator, instead of the 200+ VAC induction unit.

What I have now is a more complex system because of the three phase AC induction generator. This generator requires just the right amount of capacitance at the generator, and it requires properly sized transformers and rectifiers at each of the battery banks. The biggest problem is that neither the manufacturer, nor anyone else, could accurately specify what was needed for capacitors, transformers or rectifiers. This is highly site-specific, and in our system complicated because we are using the power at three places, each with its own transformer. I finally got the system to put out the power we needed by replacing the induction generator, capacitors and transformers with different sizes. This setup was determined experimentally. It was very frustrating, time consuming and expensive.

Our hydro system is now producing 240 watts with a ¼ inch nozzle installed at a net head of 203 feet; this works out to an overall efficiency of 36%. With the ⅜ inch nozzle installed the system produces 430 watts at a net head of 187 feet; this is an efficiency of 31%.

The Big House's PV Array

Putting trackers on the roof is an interesting design feature and a challenging installation. I first got the idea while visiting Richard Gottlieb and Carol Levin of Sunnyside Solar near Brattleboro, Vermont. They have an eight panel Zomeworks tracker mounted on their garage roof. Why put the tracker on the roof? There are three advantages for us in this application. First, it gets the PV array way up high — the top of our arrays is 32 feet above the ground. This drastically reduced the number of trees we had to clear to get the sun on the panels. And second, it saves space on the ground for other things like sandboxes and gardens. Third, we don't have to look out over the trackers from our windows. Why use trackers this far north? Usually we do not specify them here because at our latitude (45°N) they add only about 6% to the PV power production during the winter and about 22% the rest of the year. The reason we went with the Zomeworks track racks is because we have a hybrid system. I sized the PV arrays to meet all our charging needs during the summer. Our summer is a dry time, and our hydro system cannot be relied on then. The trackers add about 33% to the PVs' power production during the summer. Therefore, I reduced the total number of panels from 32 to 24 by using the trackers. The cost of the trackers was offset by the reduced cost of the downsized PV arrays.

PV Installation

Each of the two arrays above our garage roof holds 12 Kyocera 48 watt PV modules for a total of 1,152 peak watts of solar-produced power. Over the year our Kyocera panels have consistently outperformed their manufacturer's ratings. On a recent April day, I observed an array current of 42 amps at 28 VDC. This occurred on a day when the sky had many puffy, white clouds (known as cloud enhancement). On a clear sky, typically I measure 37.7 amperes charging our 24 VDC battery bank. I have an analog ammeter installed in the cover of the fused PV disconnect for quick checks. For more accuracy, I use the millivolt scale on my Fluke 23 multimeter to measure the voltage drop across the precision (0.25%) 50 millivolt shunt on our Thomson and Howe ampere-hour meter (see *Home Power* Magazine, #11, "Things that Work!" article). A 48 watt Kyocera panel is rated at 2.89 amperes, but I measured 3.14 amperes per panel.

The 24 Kyocera J-48 modules are mounted on two Zomeworks pole-mounted track racks. Each tracker was placed on its pipe mast by a crane operated by an expert and a crew of three helpers on the roof. Hiring the crane cost US$210 and was worth that and more. Installation in any other fashion would have been asking for trouble: possibly fatal damage to the PV or trackers, and/or potential injury to yours truly and my crew.

The pipe masts themselves are five-inch schedule 40 steel, each 17 feet long. The masts were cut seven feet from the base and later spliced with a four-foot section of four-inch pipe inside the five-inch pipes. The splice was necessary because the full 17 feet length would not fit into my shop easily nor would it push up through the roof easily. The lower section of the mast (7 feet) had an 18 inch by 18 inch plate of ½ inch steel welded on its bottom. The steel plate was drilled out for three ½ inch lag bolts along each side. The masts were bolted down with eight bolts per mast. The lag bolts went through the ¾ inch tongue and groove plywood decking and into the 2 × 10 floor joists and added box bridging. The upper section of the mast (10 feet) was lowered through the hole in the roof by two men to a third man guiding it into the splice insert. A standard roof flange of aluminum and rubber was then placed over the top of the pipe mast and seated onto the roof where it sits under the high shingles and over the low shingles. The seam where the roof flange and pipe meet is sealed with a type of butyl tape called Miracle Seal. This thick, pliable tape expands and contracts with the steel pipe during changes in temperature.

The last detail of the mast's installation was fastening the pipe to the roof rafter for stability. Absolute rigidity is as important here as it is at the base plate. Consult with a local building expert or structural engineer if there is any doubt about your roof-mounted tracker. We placed the masts right next to a 2 × 12 roof rafter, added shims there to tighten this union and then securely bolted the pipe to the rafter with a large steel U bolt. With the pipe fastened securely at its base and at ten feet (leaving seven feet above the roof), we met the Zomeworks installation requirement that half of the mast be buried in concrete below grade. I have witnessed wind gusts of over 55 miles per hour make the arrays flutter from side to side (buffered by the shock absorbers on the trackers), but the same gusts do not move the pipe masts at all.

We drilled a small weep hole at the very bottom of the pipes to drain condensation and prevent rusting from inside. The pipe masts were grounded for lightning protection with #4 bare copper wire at the base plates. The ground wires were bonded together with a split bolt connector to a common wire which ended in an eight-foot driven ground rod bonded to the main system ground.

The arrays were mounted on their trackers in our garage and wired in series and parallel for 24 volt operation. Module interconnections were made with #10 sunlight-resistant, two-conductor Chester Cable terminated in a junction box on each tracker. Once the arrays were in place, we came out of each junction box with #8 gauge Chester cable. We clamped the cable to the tracker for strain relief and fed it down through a hole tapped on the top of the track rack's pipe fitting. A weatherproof connector was used here. Of course, a loop of cable was used as slack before entering the pipe, to be taken up during the tracker's movement over the course of the day. The cable was then fished out of the pipe via another hole tapped at ceiling height, and a Romex connector was used here. The two cables were run to the center of the room where a junction box fed with #0 gauge copper cable awaited them. The length of each #8 gauge cable is 26

feet. The length of the #0 gauge copper cable run from the junction box, back through the house, and down to the battery is 90 feet.

Battery Bank and Big House Loads

Our storage batteries at the Big House are Trojan J-185 deep cycle, lead-acid types. We use 14 of these 185 ampere-hour batteries in a 24 volt configuration for a total of 1,295 ampere-hours (31 kilowatt-hours) of storage. In our system they are an economical choice because we normally do not cycle them below 50% of capacity. The Big House receives 4.8 kilowatt-hours per day from the hydro when the ¼ inch nozzle is being used, and eight kilowatt-hours per day with the ⅜ inch nozzle. The hydro power is often switched off at the Big House when the sun is shining, and all the power goes to the Barn and the Cherry House. The PV panels produced an average of 3.9 kilowatt-hours per day as measured during March and April of 1990 by the T&H amp-hour accumulator.

Voltage is controlled at all three of our battery banks by Enermaxer shunt regulators. I chose the Enermaxer because all of our battery banks are charged by multiple sources. The Cherry House is charged by PVs, hydro and an engine/generator. The Big House is also charged by these three sources, while the Barn is charged by hydro and engine/generator. The Enermaxer is connected to the battery bank and to shunt loads. It doesn't matter what the charging sources are as long as the current rating of the shunt loads are equivalent to the highest possible amperage of all charging sources combined at that particular battery. The Enermaxer works well because it smoothly tapers the voltage of the batteries to optimum float voltage (user adjustable to a tenth of a volt).

We average about 4.8 kilowatt-hours per day of power consumption in the Big House, with six kilowatt-hours peak during a busy, wintertime wash day. We are able to satisfy our power requirements and keep our battery bank quite full without using the generator because of our hybrid PV/microhydro system.

The loads in the Big House (14 rooms plus a full basement) are typical for a busy family of five. Various lighting products (all DC) have been used with good results including LEDs for night lights. During our long winters, we average around 140 ampere-hours or 3,360 watt-hours used on lighting per day. Other 24 VDC loads include a Sun Frost R-19 (19 cubic foot refrigerator) and a Sun Frost F-10 (10 cubic foot freezer). I recently recorded their individual power consumption on my portable T&H amp-hour meter over a test: the R-19 used 23 ampere-hours (552 watt-hours) per day. The F-10 used 28.65 ampere-hours (688 watt-hours) per day.

Our 120 VAC loads include a washing machine (350 watt-hours per use), a clothes dryer (propane fired with electric motor — 150 watt-hours per use), an automatic dishwasher (275 watt-hours per use), a stereo system, a 19 inch color TV that uses 80 watts with the VCR (65 watts alone), the controls on the Essex boiler (40 watts) and other appliances/tools. Our total average 120 VAC power consumption per day has been running around 2.5 kilowatt-hours per day.

The inverter we are using is the Trace 2024 with standby battery charger, turbo cooling fan and remote digital metering. It is able to handle the washer, dryer and dishwasher all

Total System Costs	
Wire, cables, conduit, fuses, breakers, distribution panels, disconnects, boxes, fans and all labor	$22,400
Cherry House system including generator, refrigerator, lighting, all wiring and labor	$5,600
Winco 12,500 watt generator setup	$4,500
Microhydro system includes everything except building the pond and turbine shed	$4,377
Barn system — everything included	$3,700
Big House System Specifics	
Tracked PV arrays — 24 @ Kyocera J48 modules, 2 @ Zomeworks trackers, installation etc.	$9,500
Sun Frost R-19 refrigerator and Sun Frost F-10 freezer	$3,800
DC lighting — high efficiency 24 VDC fluorescent lighting	$3,125
Battery bank — 14 @ Trojan J-185 lead-acid batteries	$2,250
Trace 2024 inverter with battery charger, turbo and remote metering	$1,650
Controls and instrumentation	$695
Grand total of all three systems including interconnection	**$61,597**
Note: All costs in US dollars	

David Palumbo

at the same time. We do the laundry during the sunny days whenever possible because the batteries are full by the afternoon and the Enermaxer would just be shunting off the power surplus. A better use of the sun's energy is cleaning our 14 loads of laundry per week!

The Big House has more than satisfied our goals for an energy efficient, comfortable and versatile home for our family and my growing alternative energy business. It has helped bring alternative energy into the mainstream in our area. Our home power system is a demonstration for those considering alternative energy as their power source. It is also an example for bankers who are hesitant about lending on non-grid-connected property. We have been able to open some eyes and get a few projects going that would otherwise have never left the drawing board.

I wish to thank those contributors I have already mentioned. Also all the fine people who worked on the project, most notably Gary Cole (Electrician), George Stone (Carpenter) and David Vissering (Jack of All Trades).[1]

Editor's Note: This article first appeared in *Home Power* Issue #17 (June/July 1990) and is reprinted with permission of the author.

8

A Visit to Hydro Oz

MALCOLM TERENCE

We drove down the winding, rutted mountain road, like Dorothy and her companions approaching Oz. We had left behind sunny Santa Cruz, California, with its beach traffic and tourists, and were creeping over the ridges into a shady, steep canyon, thick with tall redwoods.

We were approaching the home and workshop of inventor Don Harris, builder of the tireless little hydroelectric plant I'd bought more than 15 years earlier. The turbine had worked — with only a little tinkering — ever since, churning out more than 300 watts continuously at my home in Northern California.

It had delivered — let me do the math — 40 megawatt-hours and outlasted every appliance or power tool I'd ever owned. But now it was broken. We'd neglected a worn bearing too long, and the same water power we'd harnessed for years had started an oscillation that chewed up several parts in a few noisy days while we were away.

Hydro Breeder

Harris was standing in one of the few flat spots outside his home when we arrived. "Up there's where most of the work gets done," he said

with a nod, "and down there is the power plant for all our electricity. This is really a breeder facility — using hydro to build hydro." He explained that nuclear promoters in the 1950s said the United States could use reactors to

Harris' workshop, perched on a steep hillside in the Santa Cruz Mountains, is powered entirely by one of his turbines. He has produced more than 3,000 turbines here.

Malcolm Terence

Malcolm Terence

Don Harris holds a turbine on its side to show the four-nozzle layout designed for sites with high flow.

make their own fuel and solve all our energy and pollution problems at once. Somehow Harris' claim to *breeder* sustainability seems less sinister than the nuclear industry's.

Harris' turbine looks like a streamlined bread machine at the end of a two-inch PVC pipeline that snakes down out of the forest. An alternator like one in a car was bolted snugly onto its top. Hidden underneath was a small bronze waterwheel with a fringe of double cups, each the size of a tablespoon. It sat near the edge of a tiny stream channel and looked much like mine, but without the covering box I'd built to protect mine from the elements. "I leave it out in the weather and abuse it to see how it does," Harris said with a grin. "So far, so good."

The Hydro Doctor Is In

We pulled out a sack of parts from my turbine, and Harris pored over them like a patholo- gist. He carefully counted and inspected the cups on the small brass waterwheel, which is the real center of the turbine. "Fifteen cups," he said. "You've had this for some time. I switched to 17 cups to get a few percent bet- ter efficiency." He led the way up to his shop, past shelves stacked with shiny new housings and other components, and into a small room crowded with machine tools.

"I started back in 1981. Back then, I riveted the cups on the wheels. I milled the cups one at a time for the first 50 wheels. I remember I got the idea from a book called *Hydro Power for Home Use*. Nobody was building anything for small creeks. Then we switched to castings, where we build the wheels in wax and take them to a foundry where they're poured. First we used aluminum. Your wheel is the first kind we cast from bronze. Bronze was a little more resistant to erosion, and there's less elec- trolysis between the metals of the wheel and the alternator shaft."

The wheels with their distinctive double cups, Harris explained, were invented by Les- ter Pelton to win a federal design contest in 1881. Another designer, Abner Doble, refined the shape of the notch and the double elliptical catches. One of Harris' gifts is to say phrases like "double elliptical catches" to people as though they understand what that means.

Harris said that impulse turbines such as Pelton wheels, as they are still called, and Turgo wheels are much more tolerant of wear from silt in the water. The silt has a more de- structive effect on reaction-style turbines that use propellers machined to very sensitive tolerances. Then he looked grimly at the advanced wear on my wheel. "It might still have a few more years on it," he said. Deftly he ran a shaft through the center of my wheel as an arbor and tested it for balance. With the drill press he removed a little dimple of brass near the center, tested again, drilled another place, tested again and declared it fine for use.

Since he started, Harris has built more than 3,000 of his turbines for small hydro operations all over the west and as far away as Appalachia. He started with automotive Delco alternators, switching to Motorcraft alternators for big wattage gains. Recently, he began using permanent magnet alternators for even greater efficiency.

Head and Flow

Don Harris sells a range of hydro plants that go from a 12 volt, single-nozzle model at around US$1,000 to a four-nozzle, permanent magnet model that sells for US$2,020. The variables are the amount of water and the amount of water pressure.

Pressure is a function of how far vertically above the turbine water enters the pipe — what hydro people call *head*. The diameter, type and length of the pipeline also affect the pressure that can be delivered to the wheel.

One of Harris' turbines was operated at 1,000 feet of head at the Grand Canyon. It doesn't take very much water to generate some electricity at that much pressure. The nozzle was almost a pinhole (approximately 0.070 inches) to get 300 watts. When inquirers have less than 25 feet of head, Harris often refers them to Paul Cunningham of Energy Systems and Design or Ron McLeod of Nautilus Turbines because their machines are designed for

Top: This generation of Harris turbines used Motorcraft alternators. He has since switched to permanent magnet alternators for increased efficiency, durability and reduced maintenance.

Bottom: Don Harris demonstrates the final truing of a new bronze wheel on the machinist's lathe in his workshop.

Malcolm Terence

Malcolm Terence

Malcolm Terence

Bliss Kok, Harris' co-worker, carefully assembles the wax plug of a turbine wheel. The wax is cast in plaster, then melted out, after which the plaster mold is filled with bronze. A new wax pattern is used for each wheel.

the higher flows needed to compensate for low head.

Harris reeled off what he calls his basic law:

Potential output in watts equals head in feet, multiplied by flow in gallons per minute, divided by eight for a permanent magnet model or by ten for an alternator.

Inefficiencies creep into the equation, he said, but because hydro works 24 hours a day, its output needs are lower than solar or wind power to still achieve the same daily energy generated. This is especially true at my home, where the sun peeks over the ridge to the south for barely two hours a day around winter solstice.

Quite a few of Don's systems operate with relatively low flow, in the three to four gallons per minute range. But with 300 feet of head, a Harris wheel can still generate more than 100 watts (more than 2.4 kilowatt-hours per day) at these low flow rates.

Positive Anarchy

Harris interrupted his explanation to examine the next broken part in my bag, a tube that held the nozzle. A gaping hole was worn in its side from the vibration. He shook his head both at the level of wear and at the age of my unit. "We don't make them like that any more. Let me build you one and convert you to a hose fitting. It'll hold up better if you get vibration again."

Harris pulled a part out of a box on the floor and mounted it on an ancient machinist's lathe. He tugged the belt to overcome the motor's inductive load and start it in motion, and 60 seconds later he'd shaped the piece of a PVC thread/slip adapter, designed for a whole different purpose, into a near-perfect hose nipple.

"Where'd I put the hacksaw?" he asked no one in particular and said, "It's a little bit of anarchy in here," with a gesture across the cluttered shop. To Harris, anarchy is probably high praise, if not a great organizing principle for a shop.

Don told a story about installing a system in Nicaragua during the Contra War in an old Somoza-regime hacienda that had been taken over by peasants. When the American-backed Contras blew up the main utility lines nearby in 1987, the Harris system generated the only electricity in the region. Soon after, Ben Linder, one of Harris' American coworkers, was assassinated by Contras. "The Contras were terrorists backed by Reagan, and the consequences of terrorism are borne by the common people," he concluded.[1]

Vision

Harris said the switch to permanent magnet alternators delivered higher wattage output with the same flow and pressure input. On the other hand, it has more awkward controls than the Delco-type system used on my machine, which adjusts with the twist of a knob. The permanent magnets require stopping the turbine for each increment of adjustment. While he put the finishing touches on the hose fitting, Harris announced that he was working on a better permanent magnet system, one that came to him in a vision.

Speaking over the hum of the lathe, Harris told a long story that began, "Once I gave five rupees to a beggar in Nepal…" and ended several minutes later with "…on the way back across Utah, I saw the Tibetan priest again sprinkling out the dust." It's hard not to trust a vision for an invention that began in Nepal and ended in the Utah desert, so Harris has dedicated months to constructing the next generation of permanent magnet turbines, machines that can be adjusted without being shut off. He pulled a prototype out of another box of anarchy in the corner and started explaining to me about magnetic lines of flux, a topic that made me long for the simplicity of double elliptical catches.

Although his father was a physicist working in optics, Harris says he is mostly self-taught, with a few harmless forays into Southern California colleges 40 years ago. "Most of what I know about mechanics and fabrication I learned from building drag racers," he said as he buffed the hose fitting, removed it from the lathe and threaded in a ½ inch nozzle. He

---

Don Harris' Hydro System

Don's penstock (water delivery pipe) is 1,300 feet long, with three-inch PVC reducing to a pair of two-inch lines. The head (vertical drop) is 161 feet, and he uses a 5⁄16 inch nozzle delivering 21 to 22 gallons per minute at peak flow. Don is using a single nozzle for his permanent magnet alternator, the first one of the current series, which he built and installed two years ago. At peak flow, the hydro plant produces 430 watts of 24 volt electricity. In the dry season, he has seen his output fall to 80 watts. He uses a 200 foot run of #4 copper wire to carry the output to a series of 20, single-cell, nickel-iron batteries. Each is 1.2 volts and 270 amp-hours. Harris says they are inefficient and guzzle replacement water, but last forever and are environmentally benign. Besides, he says, they are easier to pack than an L-16 battery. For the workshop machinery, a six-year-old Trace SW4024 inverter gives Don AC. Battery voltage is controlled with a 20-year-old Enermaxer regulator. His house nearby still uses 12 volt DC electricity, although he also uses a DR series Trace 2512 to provide some AC. On the roof of his workshop, he has installed a 900 watt photovoltaic array, but he hasn't needed them enough yet to hook it up to his batteries.

handed it to me with a length of special hose and a replacement field adjustment rheostat.

Finally we had all the histories, repaired parts and visions that we could manage, and we bid Don a goodbye with an invitation to visit our canyon further north in California someday. I remembered in the end that Dorothy's companions also got all they needed when they visited the Wizard of Oz. I felt like a modern day scarecrow. "Let's see, I needed a brain, a ½ inch nozzle and a new rheostat. Now I'm set."[2]

Editor's Note: This article first appeared in *Home Power* Issue #99 (February and March 2004) and is reprinted with permission of the author.

Historic Oregon Trading Post— A Renewable Energy Model for the Public

JOHN BETHEA

The Rogue River Ranch, which is on the US Register of Historical Places, is located in southern Oregon on the beautiful, wild and scenic Rogue River. The US Bureau of Land Management (BLM) has managed the Ranch since 1970. Having used propane-fueled generators at the Ranch from 1970 until the present, the BLM saw an opportunity to get away from all the fuel expenses, noise, pollution and mechanical breakdowns. With the Ranch averaging about 20,000 visitors a year, what an opportunity to demonstrate renewable energy!

Learning About Hydro Power

In 1980 while living at and working full-time restoring the Rogue River Ranch, I was rummaging through and trying to organize one of the old buildings. A 1940's era cast-iron, 12 inch Pelton wheel caught my eye. After doing a lot of research and desperately wanting to free myself of the hassles associated with operating propane generators (the Ranch is about 30 miles as the crow flies from the nearest power lines) I rebuilt and hooked up the Pelton wheel. We had a gravity-fed water system with about 2,500 feet of two-inch poly pipe creating a static pressure of about 60 pounds per square inch.

I went to the local alternator shop and got a Delco that they thought would work the best. I put two 12 volt 8D batteries in parallel and buried #4/0 USE aluminum cable for a run of 200 feet to the caretaker's house. I wired the house for a 12 VDC system which included lights, TV and a small communications system. All of the major appliances were propane.

The Pelton wheel had a 17 inch pulley wheel on it, and I put a belt to the alternator resulting in the alternator putting out about five amperes. In the winter I could run the wheel as much as needed, and in the summer about 25% of the time since the water was needed for irrigation. That was plenty for me at the time. I'd only have to run the propane generator once in a rare while!

Learning about PV and Inverter Systems

The Ranch had been restored and set up as an interpretive cultural site for the history of the entire area. In a year it sees about 20,000 visitors that can take a self-guided tour of the buildings and enjoy the grounds and setting. I no longer live full-time at the Ranch but supervise the caretakers who live there from May first to November first each year.

Late in 1991, while visiting the current caretakers Laura and Loren Rush at their winter home in Baja California, I was very impressed with some of the PV/inverter systems their neighbors had. I felt that adding a PV/inverter system to the 12 VDC hydro system at the Ranch would not only fit in well but really cut back on using the propane generator. Generator power demands had again increased over the years.

Wanting to learn more, I spent a lot of time at the library where I found lots of reading material. I found the best help in Real Goods' *Solar Living Source Book* which contains a lot of easy to understand information.[1] In 1991 when a decision was made to put the money we had budgeted for a new generator into upgrading the hydro system and adding a PV/inverter system, the technical help received from the Real Goods staff was very comforting.

Hydro

We upgraded the hydro system by replacing the alternator with a custom-made one and reduced the pulley wheel on the Pelton wheel from 17 inches to 15 inches which gave a better power ratio. The water system was upgraded by replacing all the piping and getting more elevation on the intake. The system now has about 200 feet of head starting with about 800 feet of four-inch PVC then 700 feet of three-inch PVC and finally about 1,000 feet of two-inch PVC. There is a 1,500 gallon tank at the head. This resulted in increasing the static pressure to about 85 pounds per square inch with about 65 gallons per minute free flow. The hydro system now produced about 14 amps at 12 VDC or about 170 watts.

Solar

We installed six Siemens M-55 modules on a homemade frame and used an SCI 30 amp controller. We installed four new Trojan L-16 batteries, a Trace 2512 inverter, an APT 400 amp disconnect and a TriMetric meter with a 500 amp shunt. This system worked really nicely. The problems were mainly that in a few short years the demands for power had been steadily increasing, and this was a fairly small system. The Trace 2512, being a modified sine wave, would occasionally burn out something like cordless drill battery chargers! Also, the old Pelton wheel wouldn't work well in the fall when available water volume dropped. In all candor we had a few wiring and fuse installations that were less than what code calls for.

Opportunity to Get Educated

With the exception of helping put in a few other small PV systems, my experience and knowledge weren't really going anywhere. In the summer of 1995, I jumped at the opportunity to attend the Advanced Photovoltaics and Wind Power courses at Solar Energy International in Carbondale, Colorado. The four weeks I spent there confirmed a lot I had been doing right and wrong.

Partnership

Seeing the need for a much larger renewable energy system at the ranch, a proposal was made for a new system, but there wasn't much hope of getting it funded. Then in September 1995, a memorandum with a survey attached came across my desk. The Bureau of Land Management (BLM) and the US Department of Energy's Sandia National Laboratories' Photovoltaic Systems Assistance Center formed a partnership titled *Renew the Public Lands*. The purpose of this partnership is to expand the use of photovoltaics and other renewable energy sources within the BLM.

Under the partnership, a comprehensive survey of current BLM photovoltaic use and acceptance was conducted. In addition to the survey, new opportunities for the expanded use of photovoltaics were identified, and several pilot projects were developed. The Rogue River Ranch was selected as one of the pilot projects. The Medford District (BLM) agreed to provide about 65% of the funding needed, and Sandia National Laboratories provided the rest. We agreed to use Sandia's money only for construction so it would not appear they were pushing any particular product. Trent Duncan, an engineer with the BLM Utah State Office, and Hal Post with Sandia National Laboratories oversaw more than 30 projects completed in 1996 on BLM lands nationwide.

Coming up with the Bureau of Land Management's Medford District share of the cost was a problem. This was solved by doing a major trail maintenance and construction project with volunteer help instead of contracting the project out as was budgeted. This provided enough savings to fund the project. I'd like to thank the Roaring Rogue District of the Boy Scouts of America and veterans from the White City Domiciliary for their help! They not only helped with getting the trail work done but with the Renewable Energy Project as well.

Putting It All Together

The first step was to come up with a design for the system, not easy when you haven't done a lot of this before. After completing a basic design I took the liberty of asking a lot of questions from a lot of people. The technicians at

Applied Power Corporation in Lacey, Washington, where I bought a lot of the components were very helpful and knowledgeable. Don Harris, whose Pelton wheel we used, was very friendly and helpful. I even bugged Richard Perez at *Home Power* a couple of times. The technicians at Ananda Power Technologies fielded most of my calls, and I can't say enough about their willingness to help and advise.

Getting Started

It was early this winter when actual construction started. Jason Miniken who works for the Medford District BLM and myself worked on the project as often as time and weather permitted. We built an 18 by 9 foot control building to house the two hydro units, most of the electrical and electronics and the two separate battery banks. We built two completely different systems side by side in the same building.

First system — 12 volt DC

Most the components that we had in the existing 12 VDC system were taken down and used in a redesigned system. However, the Trace 2512 inverter was taken out of the system. We installed proper fused disconnects, new wiring, two 12 VDC distribution centers, a 300 watt Ananda TDR (Tapering Diversion Regulator) for the old Pelton wheel, a low voltage disconnect, a new pole-top mount for the Siemens M-55 modules and a remote meter. We cleaned up and reused the four existing Trojan L-16 batteries.

Second system — 24 volt DC

This system has 12 Trojan L-16 batteries. There are two sub-arrays, each with six Solarex MSX-83 modules. A two-nozzle Harris Hydro unit was installed. Only one hydro unit, the

The Rogue River Ranch 12 Volt System

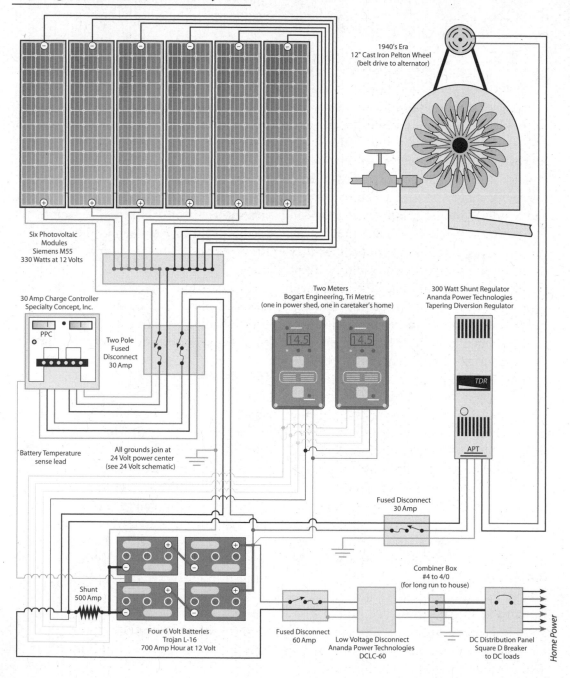

1940's Era
12" Cast Iron Pelton Wheel
(belt drive to alternator)

Six Photovoltaic
Modules
Siemens M55
330 Watts at 12 Volts

30 Amp Charge Controller
Specialty Concept, Inc.

PPC

Two Pole
Fused
Disconnect
30 Amp

Two Meters
Bogart Engineering, Tri Metric
(one in power shed, one in caretaker's home)

14.5   14.5

300 Watt Shunt Regulator
Ananda Power Technologies
Tapering Diversion Regulator

TDR

APT

Battery Temperature
sense lead

All grounds join at
24 Volt power center
(see 24 Volt schematic)

Fused Disconnect
30 Amp

Shunt
500 Amp

Four 6 Volt Batteries
Trojan L-16
700 Amp Hour at 12 Volt

Fused Disconnect
60 Amp

Low Voltage Disconnect
Ananda Power Technologies
DCLC-60

Combiner Box
#4 to 4/0
(for long run to house)

DC Distribution Panel
Square D Breaker
to DC loads

Home Power

The Rogue River Ranch 24 Volt System

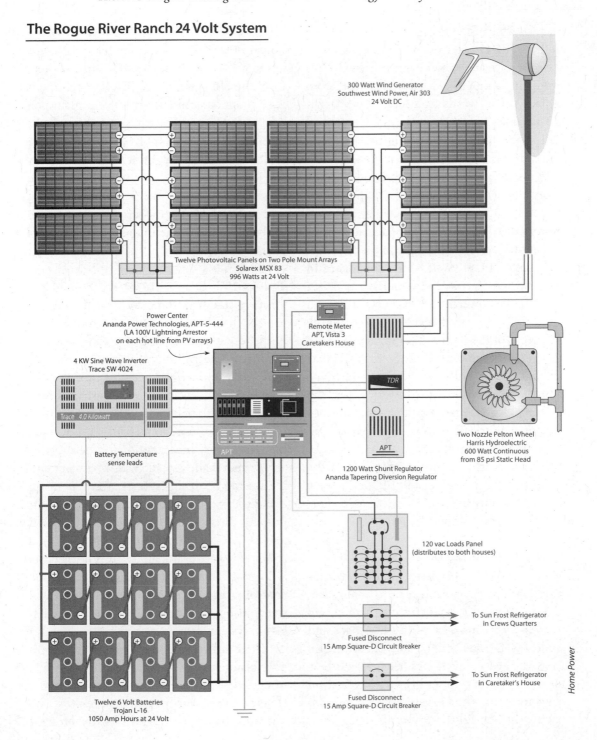

300 Watt Wind Generator
Southwest Wind Power, Air 303
24 Volt DC

Twelve Photovoltaic Panels on Two Pole Mount Arrays
Solarex MSX 83
996 Watts at 24 Volt

Power Center
Ananda Power Technologies, APT-5-444
(LA 100V Lightning Arrestor
on each hot line from PV arrays)

Remote Meter
APT, Vista 3
Caretakers House

4 KW Sine Wave Inverter
Trace SW 4024

Trace *4.0 Kilowatt*

Two Nozzle Pelton Wheel
Harris Hydroelectric
600 Watt Continuous
from 85 psi Static Head

Battery Temperature
sense leads

TDR

APT

1200 Watt Shunt Regulator
Ananda Tapering Diversion Regulator

120 vac Loads Panel
(distributes to both houses)

To Sun Frost Refrigerator
in Crews Quarters

Fused Disconnect
15 Amp Square-D Circuit Breaker

To Sun Frost Refrigerator
in Caretaker's House

Fused Disconnect
15 Amp Square-D Circuit Breaker

Twelve 6 Volt Batteries
Trojan L-16
1050 Amp Hours at 24 Volt

Home Power

Harris or the old cast-iron Pelton wheel that's in the 12 VDC system, can be run at a time as they use the same water source. The Harris is producing about 25 amps at 24 VDC, or about 600 watts. The wind turbine is an Air 303 that we put on a 27 foot tower beside the control building. Admittedly it's not the best wind site and the tower isn't very high, but we wanted to demonstrate wind power. We were concerned about safety and aesthetics of a higher tower. The Air 303 puts out one to two amps a few hours a day. We haven't really had much wind yet, but we expect wintertime to produce better results. A 1,200 watt APT-TDR is used to regulate the Harris Hydro and Air 303. An APT 5-444 Powercenter was installed and a Trace 4024 sine wave inverter powers all the AC needs of the Ranch. Two Sun Frost RF-16 refrigerators are run off of the 24 VDC battery bank. A 35 amp low battery cut-out in the power center protects the batteries from the Sun Frost loads. That's also a very good incentive for the caretakers to keep the batteries at their proper charge level!

Other than fine tuning, we finished the project the first week of July, 1996. The new system has only been up and running a short time, but so far it's doing nicely.

Why Have the 12 Volt DC System?

Most of the 12 VDC wiring and light fixtures were already in. We did replace some of it, and the distribution centers were brought up to code. There are a lot of 12 VDC loads and some, like the communications center, are on 24 hours every day. It was felt it would be a lot more efficient not to have the inverter on constantly. The same reasoning was applied to having the Sun Frost refrigerators at 24 VDC

vs 120 VAC. Plus, what if the inverter broke down? There is also a completely separate lighting system with the 12 VDC that was kept from the old system. The 12 VDC lighting is used the most and takes a lot of pressure off the 24 VDC system by not going through the inverter. There are small 12 VDC lights above the beds to read at night. It wouldn't be very efficient to have to have an inverter on just to power them. If most the wiring and fixtures for the 12 VDC system weren't already in I'd probably not put one in again, but I'm sure glad it's there!

What the Two Systems are Powering

Power is supplied to most of the buildings at the ranch. The caretaker's house, the main house, crew quarters and shop have most of the loads. The communications system, fuel tank pump and irrigation system are all powered by the renewable energy system. I estimate that the appliances powered by the 12 volt system consume about 850 watt-hours per day. The energy consumption for the larger 24 volt system is about 6,300 watt-hours per day.

The ranch's extensive irrigation system operates every other hour for a total of 12 hours per day. This system is powered by 12 VDC and 24 VAC (via the inverter) and consumes about 96 watt-hours daily. A Photocomm SIPS controller, a Hardie Irrigation TC-2400 controller and eleven Weathermatic 8000 CR solenoid-operated valves are used in this irrigation system.

There is a large three phase 240 VAC propane generator to operate a centrifugal pump in case of extreme firefighting needs and a smaller 4,000 watt propane generator as a domestic backup. My guess is the propane

Other BLM Renewable Energy Projects

The BLM administers what's left of the US' once vast land holding that has not been passed on to other individuals, industries, states or federal agencies. This amounts to over 272 million acres. It also manages mineral estate under an additional 300 million acres that are owned or administered by other agencies or private interests.

PV has been used for many years at the remote BLM facilities, but it wasn't until April 1995 that the *Renew the Public Lands* partnership was forged with Sandia National Laboratories' Photovoltaics Systems Assistance Centrer. The goals were to survey existing PV uses, identify barriers to expanded use and identify potential new opportunities within the BLM. Partnership cost share funds were made available for pilot projects that would expand BLM's familiarity and experience with PV technology. Here's a list of projects in addition to the Rogue River Ranch Project:

- Sand Wash Ranger Station, on the Green River, Desolation Canyon, Utah: Electrify residence and 12 VDC communications at the contact station. 1.4 kilowatt array, 3,600 amp-hour battery, 1,500 watt inverter and propane backup generator.
- Ward Jarman's South Camp Cabin, Book Cliffs, Utah: Electrify remote administrative site. 330 watt array, 530 amp-hour battery, 1,500 watt inverter and propane backup generator.
- Kane Gulch Visitor Contact Station, Cedar Mesa in San Juan County, Utah: Electrify visitor station. 1.4 kilowatt tracking array, 3,600 amp-hour battery, four kilowatt sine wave inverter, propane backup generator.

Batteries are in an underground concrete vault to help reduce performance impact from extreme temperatures.
- Hickison Petroglyphs, near Austin Texas: Provide drinking water from existing well to visitors. System to inclue a PV powered pump, no specs available.
- Burro Creek Campground, near Kingman Arizona: Light restrooms and pump water with PV, provide electricity for campground host with portable PV system. No specs available.
- Hobo Camp, Westwood, California (near Susanville): Portable PV system for host camp. 380 watt array, 480 amp-hour battery, 800 watt inverter, all on a trailer.
- Mine Shaft Spring Butte District, Montana PV: Power pumps water from mine shaft to storage tank and stock troughs. No specs available.
- Portable PV systems, 13 scattered through Arizona, California, Colorado, New Mexico, Oregon and Utah. One kilowatt-hour per day for basic AC electrical needs. No further specs available.
- Powder River Basin, Casper District, Wyoming: Early warning system for possible adverse effects of coalbed methane production, data logging at seven stations. Small PV systems, no other specs available. Nine more stations planned.
- Cottonwood Creek, Natrona County, Wyoming: Watershed monitoring of grazing impacts, powering data loggers and radiotelemetry. Eight small PV systems, no specs available.

Rogue River Ranch RE System Upgrade Cost

Item	Cost	Item	Cost
12 – Solarex MSX-83 PV modules	$5,412	1 – TriMetric meter	$122
2 – Sun Frost RF-12 refrigerators/ freezers	$4,306	1 – 60A two-pole fused disconnect 12 volts	$120
1 – Trace SW4024 inverter	$2,331	13 – Battery cables	$104
Cable and wire	$2,179	2 – Trace BC10 inverter cables	$101
12 – Trojan L-16 batteries	$2,091	2 – 15 amp breakers & boxes at Sun Frosts	$90
1 – Power shed construction	$2,000	2 – 30 amp two-pole fused disconnects	$75
Conduit, connectors, hardware	$1,500	1 – Vista 3 meter	$74
1 – Ananda Power Center	$1,178	1 – LBCO35 35 amp low voltage disconnect	$74
1 – Harris Hydro	$1,000	2 – LA100V lightning arrestors	$70
27 – Miscellaneous low voltage lights	$542	2 – Trace BC5 Inverter Cables	$60
1 – Air 303 wind generator	$441	1 – Combiner box 2@#4 to 2@#4/0	$60
1 – APT -TDR1224A regulator	$431	1 – Conduit box for SW4024 inverter	$53
2 – Pole mounts for MSX-83 PV modules	$400	1 – APT-ACS AC section for APT P/C	$52
1 – APT-TDR312A regulator 12 volt hydro	$236	2 – LB15 15 amp breakers - Sun Frosts P/C	$44
16 – 15 amp DC circuit breakers	$225	1 – RMBT remote meter terminal block	$34
1 – Pole mount for M55 PV modules	$217	1 – IB60 60 amp breaker for hydro/wind	$27
1 – APT-DCLC60 60 amp disconnect 12 volts	$211	1 – ACB60 60 amp breaker for AC loads	$27
1 – Alternator for 12 VDC hydro $200		1 – BCT-10 battery temp sensor	$22
3 – Combiner boxes for subarrays	$180	1 – A60P-30 amp input breaker $21	
2 – 12 VDC distribution boxes	$180	**Total**	**$26,788**
6 – Polaris IT250 connector blocks	$168	*Note: All costs in US dollars*	
1 – Vista 3-SH meter with 2 shunts	$130		

John Bethea

generators will only be run every once in a while just to keep them lubricated. Just what I've been wanting all along!

Comments from Laura and Loren Rush

This is our fourth year as volunteers at the Ranch. Our new renewable power system has greatly improved our comfort and ability to enjoy some of the creature comforts we couldn't have before. One of the great enjoyments is the ability to run our fans for cooling during the hot summers. The new Sun Frost refrigerators allow us to keep vegetables for at least a week longer. We only shop once a month so this is a great benefit to us. It is great not to have to run noisy generators or worry about consuming fuel, constant maintenance and repair. We are using appliances including the automatic washer as if we lived in the city. It's a great boon to isolated country dwellers.[2]

Editor's Note: This article first appeared in *Home Power* Issue #55 (October–November 1996) and is reprinted with permission of the author.

10

A Working Microhydro
at Journey's End Forest Ranch

HARRY O. RAKFELDT

We make our own electricity with a micro-hydro power system. When we were looking for our acreage, our list of requirements contained self-sufficiency. Surface water was a prime ingredient on our list. And we found it. The project to design and install our micro-hydro power system spanned four years. Our goal: to live in a *normal* electrical way, without any commercial power.

Setting the Scene

Our homesite, at 4,300 feet elevation, is located on a corner of a half-mile-wide, 80 acre, steep mountain property. We are located about one mile from commercial electricity. One of the two year-round creeks (really a stream) enters our property at the NE corner from the BLM (US Bureau of Land Management) land behind us and flows SSW across our land for about 1,800 feet. From top to bottom there is a total head of 300 feet. The creek's average seasonal flow varies between about 34 to 50 gallons per minute. But during heavy rains and snow melt, flow will go well above 100 gallons per minute. For practical hydro purposes, it is a *low* flow, *high* head site.

Our Considerations

- We like our creature comforts. We wanted our new home to be in all appearances the same as Dick and Jane's in the city.
- Because our maximum output would be low this meant a mixture of electric and propane appliances to reduce electrical needs.
- Our stream flow is heavier in the winter when needed the most.
- To produce a respectable output, the turbine would have to be located at some distance from the homesite. Thus, line loss from transmission of low voltage would be a factor.
- Output from the turbine would not meet *peak consumption* (maximum amount of electrical energy needed at any one time). To meet peak consumption, a battery bank and inverter would be required.
- The system should meet our need for *total consumption* (the number of kilowatt-hours used in a given period of time, most commonly kilowatt-hours per month).
- And money…How much would a system cost? What compromises did we have

to make? There wasn't going to be any money for a second shot if the first try didn't score — we were going to build a home at the same time. And this made me nervous.

- To make a major decision such as this about which I only had "book" exposure put me on the spot with my wife and the few others who knew what was being attempted. With respect to this hydro thing, I felt something like a paraphrased Truman quote, "The flow stops here."

Research and Design

During the four years until our house was built, I had a number of opportunities to observe the creek. Flow was measured a number of times. On this small creek, measuring was simple — build a small dam and time the overflow into a five-gallon bucket.

I measured potential head to three different turbine sites on the creek, three times each with two different sighting levels. Starting at the lowest point considered as a potential turbine site, I worked up to the proposed intake site, recording along the way the number of times I sighted through the level and then climbed to that point to sight again. The total figure was multiplied by the five-foot-six-inch distance from the ground to my eye level to arrive at the total head. Using this method, the final spot decided on for the turbine measured out at 103.5 feet of head. And the site selected offered a fairly straight line for the majority of the penstock's length from intake to turbine and generally followed the creek's SSW direction.

In reading material related to hydro, I came across a number of potential suppliers of hydro equipment and systems. I made contact with one of these firms because the system seemed reasonable in price, was small but looked well made and offered site-selected options. I discussed with Ross Burkhardt of Burkhardt Turbines the variables — flow and head. Ross and his partner John Takes did much to help me select a system. Ross has a computer program which predicts outputs on the systems he sells. We plugged in my variables and came up with a set of predictions for a 12 volt system. Then as we fine tuned the variables (different flows and different heads), the 24 volt system evolved. What followed at a rapid pace were decisions on an inverter (to match the 24 volt output), batteries, transmission cable and other related supplies. The size of the penstock — three-inch PVC pipe — had already been a factor in the discussions with Ross and used in his computer predictions. This size presented a compromise between head loss due to friction over such a long distance — 740 feet — and a nominal size for later expansion if I wanted to extend the penstock further downhill for increased output. I planned for and incorporated this option into the way I laid out the penstock.

The System

Our hydro power system consists of an impulse-driven alternator that produces direct current (DC) to maintain a battery bank. Twenty-four volts DC is changed by an inverter to 117 volt alternating current (AC) that is passed into the home's electrical circuits through the distribution panel.

For the power plant, a Harris turbine system was bought from Burkhardt Turbines.

It is a vertical axis, 24 volt DC Pelton wheel generating setup. A 37 amp Delco alternator modified for 24 volt output is mounted on an aluminum housing and is direct-coupled through the housing to a silicon bronze Pelton wheel. My setup has two jets (one to four jets can be ordered, depending on your water flow—a site-designed option). These jets hold Rainbird™ nozzles which are available in a number of different-sized openings. My system also included a Photron voltage regulator, a 500 watt, 24 volt water heating element, a rheostat control to adjust power output at the turbine, a heat sink mounted diode (to control voltage flow direction), a panel with dual meters (volts and amps), an extra alternator and detailed instructions.

The battery bank is made up of eight Trojan J-250, six-volt, 250 ampere-hour units. These batteries are true deep cycle—listed by Trojan as "Motive Power-Deep Cycle." The batteries are wired in a series of four to develop 24 volts and then paralleled to double their ampere-hour capacity for a total of 500 amp-hours storage.

A model HF24-2500SXW inverter from Heart Interface changes the 24 volt DC from the batteries to 117 volt AC for use in the home. This inverter is wired directly into the home's electrical panel. The inverter was selected for its high surge capacity—needed for our induction motors: water pump, refrigerator and washer—and a built-in 40 amp battery charger. When connected to an AC generator, the inverter operates as an automatic battery charger while transferring all the loads to the incoming AC power. We keep a 4,000 watt AC gas generator on standby, and we have to use it once in a while.

Getting It Together

The hardest part of putting the system together was the penstock. Not that it was technically difficult, but labor and time intensive. It starts above ground from the intake barrel alongside the creek. About 40 feet later it enters the ground, a very rocky area that proved somewhat slow and difficult to dig with the backhoe. Shortly after this point, it takes a 45° turn to the right (through an elbow) and continues for some distance underground before exiting to cross above a spring's streambed. On the other side of the streambed, it goes deep underground, up to six feet at one location, to maintain grade and follows a straight course for several hundred feet. Then it takes a rapid drop down a 30% grade before relaxing its descent. About 60 feet later it makes a 90° turn to the left through two 45° elbows spaced four feet apart to reduce the sharp transition. The 90° turning point here is intentional. It allows the option to continue the penstock downhill at a later date, giving more head for increased power at a new turbine site. The 90° turn would be eliminated to allow the penstock to continue in a straight line to the new site.

After this turn, the penstock exits the ground again and plunges down an embankment 40 feet toward the creek. At the bottom of the embankment, there's another 45° elbow to level out the penstock before it enters the powerhouse.

Digging the trench took a day and a half in itself. Then the PVC pipe was placed above the trench on crossboards and carefully cemented together and left to dry for a full day before it was gently lowered into the trench and covered.

Harry Rakfeldt

A barrel on the penstock keeps the system free of silt and dirt.

At the intake end of the penstock is a 55 gallon polyethylene drum. It is connected to the small dam via a four-inch drainpipe. This barrel is used as the intake because it:

- filters the debris not trapped behind the dam.
- prevents turbulent water from entering the penstock.
- allows the sediment to settle out.
- can be located as needed with respect to the dam and penstock.
- is easy to work with.
- will last for a very long time.

When I put the connections together, I arranged the air vent and gate valve assembly so that it could be removed from the barrel and penstock easily. At the barrel the PVC pipe is threaded into the barrel, and a collar is threaded onto the coupling inside the barrel. The short section of pipe on front of the air vent is only slip-fitted into the penstock. Because I only have a low flow stream to work with, building a small dam was straightforward. The end of the drainpipe that extends into the dammed water is also protected with a trash collector made of screening. At the other end of the penstock is the simple powerhouse.

The powerhouse sits directly over the streambed on railroad ties. There is easy access to the turbine components via a removable roof. It's here I really got a chance to be creative; I even used a kitchen sink! It makes a great base to mount the turbine, permitting much easier access to the Pelton wheel and pipe connections. Laying the transmission cable wasn't difficult but required some *engineering*. The terrain from the homesite to the powerhouse falls steeply downhill. The cable was buried from the house to within 45 feet of the powerhouse in a channel dug with the backhoe. The aluminum cable I chose for the transmission line between the powerhouse and homesite is very large—#4/0 (½ inch diameter plus insulation). It came on a 1,000 foot spool and was heavy.

I placed a long pipe through the cable spool and lifted this combination onto the back of my pickup truck with the backhoe. The pipe rested on the top of the pickup bed sides and was prevented from rolling off. The truck was parked alongside the house, facing uphill. I then grabbed the end of the cable and walked it downhill, unrolling the cable easily from the elevated spool. When I retraced my steps from the powerhouse back to the home-site, I sprayed this section of the cable every 10–15 feet with red spray paint to denote this leg as the *positive* side of the line. At the truck I cut the cable and then unrolled the second leg of the pair. The length of each leg is 451 feet.

The final step was to install the compo-nents at the homesite. We had planned for the equipment by having our building pad cut into a stepped pad with a bulldozer. This re-sulted in a generous 54 inch crawl space across the front half of the home where the inverter, control panel and batteries are kept.

Because of the good instructions, the components went in "by the numbers." One of the items connected was the 500 watt water-heating element. It serves to use the excess output from the turbine. *Excess* is the electric-ity not needed when the battery bank is fully charged. The voltage regulator senses the state of charge on the batteries and when the bat-teries are full, it diverts the continuously in-coming power from the turbine to a *dump*. In this case, the dump is a water heating element immersed in a five-gallon bucket filled with water. An air heating element could be substi-tuted for the water heating element.

I didn't think I would have a great deal of excess power to dump, so I chose the five-gallon bucket initially. While I was getting a feel for the way the system performed, I could always go to a larger container of water to hold the heating element. I'm still using the five-gallon bucket.

It's a Turn On

Finally. After many hours of research, long hours of planning and double and triple-checked installation, the day came to try

Top: The Harris turbine at home in the kitchen sink. Note the loading control for the alternator on the left and the valve to shut off the water to the second jet.

Bottom: The underside of the turbine and sink shows the turbine's cups.

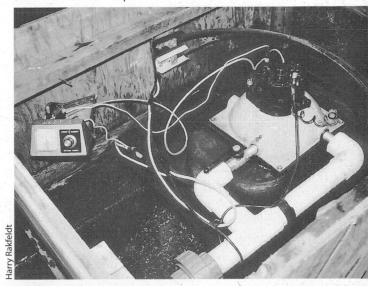

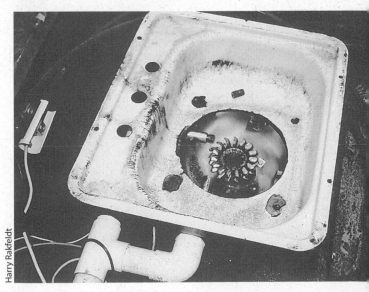

Harry Rakfeldt

Block Diagram of the Microhydro System

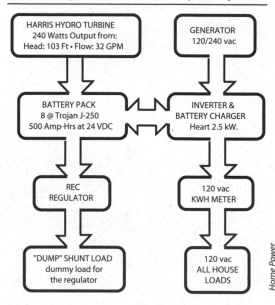

HARRIS HYDRO TURBINE
240 Watts Output from:
Head: 103 Ft · Flow: 32 GPM

GENERATOR
120/240 vac

BATTERY PACK
8 @ Trojan-J-250
500 Amp-Hrs at 24 VDC

**INVERTER &
BATTERY CHARGER**
Heart 2.5 kW.

**REC
REGULATOR**

**120 vac
KWH METER**

"DUMP" SHUNT LOAD
dummy load for
the regulator

**120 vac
ALL HOUSE
LOADS**

Home Power

out the system. The gate valve at the power-house was closed. At the intake site, I opened the gate valve to let water into the penstock. It took some minutes to fill and let air inside work its way out through the opened air vent. Then back to the powerhouse. There I slowly opened the gate valve and after some hissing and belching, the water began to flow steadily. As I continued opening the valve the turbine picked up speed and then suddenly dropped off slightly—but at the same instant the amp meter began to climb! I continued to open the gate valve and brought the system up to full output. It's working, it's working!

And for me it was a special thrill to know I had just crossed into the world of renewable energy—from and because of my resources!

That was early October 1985. Except for a period in November 1986, when I purposely shut down the system to have a modification

made to our inverter by Heart Interface, our microhydro power system has been running continuously.

Our Normal Home

It's a modified saltbox design that originally appeared as a cabin style post-and-beam plan in *Home* magazine. It's now a passive home with 1,435 square feet, six-inch walls, required insulation, two baths, two bedrooms, wood-stove heat and nine-foot-high thermal mass (brick) in the woodstove alcove.

Propane is used for the range/oven, hot water heater and clothes dryer. One hundred and seventeen volt single phase electricity is used for an 18 cubic foot, self-defrosting refrigerator (4.3 amps); ⅓ horsepower jet pump on the water pressure system (8.3 amps); clothes washer (9.6 amps); 500 watt ignitor on the dryer; ignitors on the range/oven and electric motor to turn the dryer. We also have or use AM/FM stereo, AM/FM portable radio, 19 inch color TV, VCR, typewriter, desktop calculator, 1,200 watt hair dryer, small TI computer, vacuum cleaner (3.2 amps), electric broom, Dremel hand tool, electric stapler, 500 watt slide projector, electronic flash unit, small black and white TV (tube type), electric mixer, four-cup coffee maker, 30 cup coffee pot, electric griddle, blender, waffle iron, hand iron, electric knife, ⅜ inch electric drill, tape deck, Skil saw (10 amps), ceiling fan, electric clock, battery charger (portable), range hood, soldering gun, our special radio phone, electric meter and lights.

For lights we have fixtures in the dining room (300 watts), downstairs bath (240 watts) and a 480 watt guzzler in the master bath. Our light inventory is rounded out with: two two-tube, four-foot fluorescents; one two-tube, two-foot fluorescent, a PL-type (small twin tube) fluorescent (nine watts + ballast) and various single lamp, varied wattage incandescents.

The Need to Estimate

When I was researching a system design, I kept coming across the statement that in order to develop a properly-sized system, I had to *estimate* my projected usage. Now, for those of us who are coming from a *just-throw-the-switch* type of public power environment, to estimate our usage is difficult at best. Just how much does a refrigerator run in a 24 hour period? How long do I use lights while shaving on a winter's morn? How long...And the list goes on and on.

But now I can give you some real help... because I kept track of *actual* electrical usage and *patterns* of usage with a commercial kilowatt-hour power meter wired to the home's main panel. But before we look at what has been used, let's look at what I had to work with. Total head is 103.5 feet, and dynamic water pressure at the powerhouse is 46 pounds per square inch.

In the summer, I use one ⅜ inch diameter nozzle in the turbine. This nozzle runs about 32 gallons of water through the turbine per minute. This results in nine amperes at 24 VDC or 216 watts turbine output. This amounts to about 5.1 kilowatt-hours of electricity produced daily. In the winter, increased stream flow allows me to use two nozzles ⁵⁄₁₆ inch in diameter. These nozzles run about 45 gallons per minute of water through the turbine. This ups the turbine's output to 12 amperes at 24 VDC or about 6.9 kilowatt-hours daily.

The interior of the homestead, looking south into the Siskiyous.

In the 916 days that the system has been running, we have consumed an average of 4.32 kilowatt-hours per day as measured by the kilowatt-hour meter. The system produces a daily average of about 5.0 kilowatt-hours of usable electricity once inefficiencies in the batteries, inverter, power transmission and other factors are considered. The main thing to be noted is that there isn't a whole lot of leeway. There isn't much excess electricity to worry about.

Even though our turbine output in the summer is lower, so is our average daily consumption. We're not using lights as much, may not be watching TV or using the VCR as often and clothes can now be hung on the line to dry rather than tumbled in the dryer. These all help to cut a little off our usage.

In the wintertime, or any time for that matter, we have formed the habit of not leaving lights on indiscriminately. When we leave a room, off go the lights. But we don't walk around in a blackout either. We just watch our consumption through closer attention to usage. We improved over the first months after moving into the house. And now I think we have ourselves trained.

An area that we must watch is how much load we put on the inverter at one time. When you compute the watts used by the washer, water pump and refrigerator (117 volts times amps equals watts), the total exceeds the rated output of the inverter (inverter equals 2,500 watts; combined usage of items equals 2,597 watts). When using the washer and water pump, we could turn off the refrigerator. But we don't have to. The inverter surge capacity, so far, covers us when all three of the items happen to be on at the same time. So we do our washing during the daytime when lights aren't needed. And we only use the dryer after the washing is done. The surge capacity of our inverter permits it to operate for a period of time even though the normally-rated load has been exceeded. The *length* of time that the inverter will continue to operate is directly related to the *amount* the load exceeds rating. This may

The batteries, inverter, regulator and dummy load are all housed in the crawl space under the house.

Harry Rakfeldt

be minutes to only several seconds. The surge capacity for us was a must — and well worth the few extra dollars.

Standby Power

Yes, we've had to use our gas generator backup. Especially when we have guests who aren't *trained* like we are. Lights left on in the bathrooms; hair dryers going much more often; more flushing of the toilets (our captive air tank has a 36 gallon capacity but reaches its automatic turn on when 11 gallons have been used) — just plain more use in a short time frame. Fortunately, our guest stays have not been too long — but they are noticed with respect to the system.

When our system reaches its low point of 21.9 volts in the batteries, it self-shuts down to prevent damage to the batteries. Even a few minutes wait will sometimes bring the batteries back to a safe limit, and the inverter can be reset without resorting to the AC generator. But if the load on the system at the time it shut down is high, I usually choose to start the AC generator and run it for a while to boost the batteries enough to meet the need. As our desire to use more power increases, our next move will be to increase our microhydro's output. The efficiency of my system — as it operates today — ranges from 30% to 38%. Not very good. But I knew this in advance because the Delco alternator doesn't reach its efficiency in the 24 volt output until it is used at a much higher head. Because of my low stream flow, I have only one way to go — increase head for more output.

I planned for a future increase in head with the manner in which the penstock was installed. I've replaced the first voltage regulator with one much more powerful. The Photron regulator that came with the system had only a 15 amp capacity. The new regulator has a 40 amp capacity, and the float voltage level can be user-adjusted. This new regulator is made by Renewable Energy Controls, owned by Ross Burkhardt. Ross sold out his interest in Burkhardt Turbines to his former partner, John Takes.

What It All Cost

The total cost of the system has been US$5,421.37 to date. The microhydro has been operational for 916 days and during that period has generated 4,671 kilowatt-hours of electricity. At this point in time, this calculates to an electricity cost of US$1.16 per kilowatt-hour. Over the ten year expected lifetime of this system, the electricity should cost about US$0.29 per kilowatt-hour.

Now, consider that the local commercial utility (PP&L) wanted US$5.35 per foot to install one mile of line to our homesite. This amounts to over US$28,000 for the privilege of paying a monthly power bill. The money we've spent on our microhydro system is less than 20% of what the power company wanted just to hook us up!

Some Comments on Components

PVC Pipe — Easiest to use for the penstock. It has a very low head loss due to friction. Take time to cement the sections together — and to let the cement dry properly. Originally, I tried a 90° PVC curved elbow used in electrical conduit. It didn't mate properly and blew off quite easily when the system was turned on. Had to shut down for a day to repair with the two 45° elbows.

Batteries — The J-250's I'm using don't allow too much storage capacity in my situation. The next sized battery, the L-16, has 40% more storage capacity. As I expand my system and it becomes time for me to replace my present battery bank, I plan to upgrade to the Trojan L-16W.

Inverter — For those who haven't used one before, there is some adjustment necessary. For the most part, forget using the AM portion of your AC-powered radio. The hum from the lines overshadows all but the strongest stations. Stereo and video equipment may also hum depending on make and type.

Battery cables — Have all connections soldered. My cables came unsoldered. For a while they worked fine. Then deep into the first winter I begin noticing lights blinking especially when a large appliance was on. The blinking disappeared after the cables were soldered.

Voltage regulator — This is an essential piece of equipment in a microhydro system. It will sense the correct voltage level needed to properly bring your batteries up to charge and then maintain them there. Without a regulator you'd have to personally monitor the system and then either shut off the turbine when the batteries are full or flip a switch to shunt off the excess electrical output not needed for the fully charged batteries.

Faith — Place faith in a reputable dealer. Dealers have feedback from all sorts of installations. They continue to stay in business by knowing what is happening.

System Costs

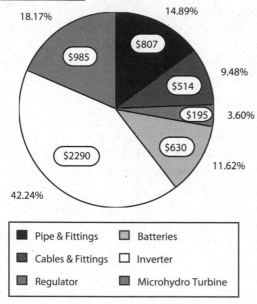

18.17%   14.89%

$807

$985

9.48%

$514

$195   3.60%

$630

$2290

11.62%

42.24%

- ■ Pipe & Fittings   ■ Batteries
- ■ Cables & Fittings   □ Inverter
- ■ Regulator   ■ Microhydro Turbine

Home Power

Closing Thoughts

First: We feel like a *normal* household. Nothing has drastically changed in the way we live.

Second: Although the list of electrical items mentioned earlier sounds impressive, we don't use many of these at any given time or the larger ones for any length of time.

Third: For the two of us, we have what we need. We can curl up in front of the VCR for a double feature, fill our 80 gallon bathtub (meaning, that every 11 gallons the water pump comes on) and other things without the system shutting down. We are careful but not fanatical about our usage.

Fourth: We made some adjustments that are now habits.

Fifth: It's not perfect. The system does work well. And so can yours. Do research, consult with distributors and have faith that you can do it too![1]

Editor's note: This article first appeared in *Home Power* Issue #6 (August/September 1988) and is reprinted with permission of the author. The original editors wrote:

When we visited Harry Rakfeldt to take the photos you see here, he had just finished moving his powerhouse some 50 feet lower than described in his article. While this change is too new to give much data yet, turbine performance has increased. The dynamic pressure at the powerhouse is now 76 pounds per square inch. The turbine's output has increased some 50% with no increase in water consumption. Harry is now considering a big-time electric hot water heater to use his additional energy.

Living With Lil Otto or
Microhydro in Seasonally Wet Climates

HUGH SPENCER

We recently purchased a Lil Otto microhydro system from Bob-O Schultze (Electron Connection) in northern California to provide us with auxiliary power during the wet season. It is probably the only Lil Otto in Australia!

Why Microhydro?

We operate a research station in the coastal tropical rainforest of far north Queensland

(in the Daintree region). Here we have no grid power and a monsoon-driven wet season which lasts from January to May, when the sky can be continually grey and the rain comes down in buckets (average rainfall is 163 inches, 118 inches of which falls during the wet).

Our home doubles as an eating area and office for the Cape Tribulation Tropical Research Station, operated by the Australian Tropical Research Foundation. Behind it is a 100 foot high gully with a stream that only flows during the wet. At the top is a small permanent spring which provides our drinking water. During the wet, this gully fairly cascades with water…

Getting Started

We ran a 525 foot length of 1½ inch black polyethylene irrigation pipe up to a small wooden dam across the gully head. This is held in place by some convenient boulders and a large cluster fig tree, obviating the need for cement which is anathema in a World Heritage area such as this. The dam is lined with several layers of black polythene to control leaks through the boulder layer that comprises the soil of the area. Several loose layers of galvanized ¼ inch mesh function as an efficient trash filter around the pipe entrance, which is itself protected by a closed tube of the same material.

The pipe snakes its way down the gully to end at the house. We didn't read Bob-O's recommendation to use white PVC piping (it has far lower friction) and paint it, until too late. At least we have an intact run of irrigation pipe to use elsewhere should we move — if the white-tailed rats don't eat it first! (These enormous native rats have a penchant for poly pipe, especially pipe smaller than 1½ inches; it gets perforated in short order.)

At the house we have a control gate valve purely to turn the water off to unblock or change jets. From this a further six feet length of pipe is connected to a modified compression fitting into which the short length of PVC pipe, supplied with Lil Otto and which carries the jet, is a jam-fit. This makes unblocking the jet a quick process. The modifications to the compression fitting are tapers cut inside the nipples of the fitting with a lathe to create smooth conical transitions from the 1½ inch to the ¾ inch jet pipe. This eliminates turbulence caused by the square edges of the pipes which can cause considerable energy losses.

Cross section of the poly adapter showing where tapers have been cut with a lathe to eliminate turbulence forming steps in the interior of the pipe.

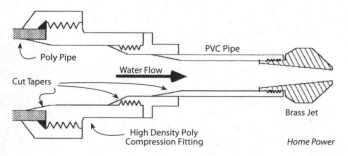

Since the present installation will probably be a temporary one, we did not want to make a permanent cement pad and drain for Lil Otto. We devised a cement block containing a wash chamber and drain, into which Lil Otto wedged nicely, eliminating the need for any clamps. It also keeps the unit stable and secure against the hydro line. We modelled a Lil Otto Body pipe-bowl-like cavity out of polystyrene foam in a box and poured cement over it. The foam was removed by digging it out and finally burning it out with a gas torch.

Subsequently, we have added a pressure gauge salvaged from an old piece of scientific gear and conveniently calibrated in kilograms per square centimeter — which means that we have to convert to kilo-pascals or pounds per square inch to make readings understandable to everyone else in the non-Imperial world (one kilogram per square centimeter equals 14 pounds per square inch). The gauge reads to 3.0 kilograms per square centimeter, which just happens to nicely encompass the pressure range of our system (2.7 kilograms per square centimeter). The pressure gauge is installed just upstream of the shut-off valve. The pressure gauge greatly assists in choosing the right jet size for the flow or observing that the water flow is lessening, as a fall in pressure indicates that the jet is too large for the available flow.

How It Works

With our 100 foot head, we get 30 watts from Lil Otto using the biggest jet provided (¼ inch diameter, 9.5 US gallons per minute), and we haven't really tried to calculate the system efficiency as there is more energy available than we can actually use. I do not use any special voltage regulator. The 12 volt hydro wiring is in parallel with the solar panels, and the homemade shunt regulator deals with both. You certainly would not want to use a series regulator or a pulse modulated regulator with a DC hydro system like this, unless you like replacing generator bearings.

Lil Otto is a vertical shaft microhydro using a very ordinary permanent magnet DC motor as the generator. It uses an injection-moulded plastic impeller, and the entire device is fitted in a housing made out of high density PVC plumbing fittings 12 inches high and seven inches in diameter at the base. It has an integral ammeter with a very useful six amp range. With its two power connection screws coming out the top like stubby antennas, the whole unit looks like a little white Dalek. I put a yogurt bucket over it to keep the top dry. A short length of ¾ inch PVC pipe fits into a collar glued (rather weakly, in ours) to the body of the hydro to hold and align the jet with the Pelton wheel. With our cement base, I found that it was more convenient to use the (now detached) plastic sleeve as a locator in the hole on the side of the unit and rely on the heavy hydro line and the concrete block to hold the alignment. In fact you can *tune* the power output by slightly rotating the body in relation to the jet. This arrangement allows one to readily

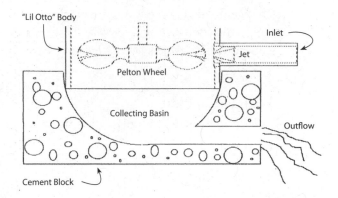

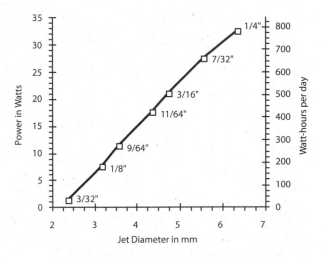

Top: Cross section of the cement block with the collecting basin and drain outflow. Lil Otto wedges into the sides of the basin and is held firmly in place.

Middle: Power (watts) versus jet diameter for a 38 pounds per square inch (approximately 100 feet height) dynamic pressure. The relationship is linear.

Bottom: Flow rate versus jet size at 38 pounds per square inch. This hydro is designed for low flows — less than 16 gallons per minute.

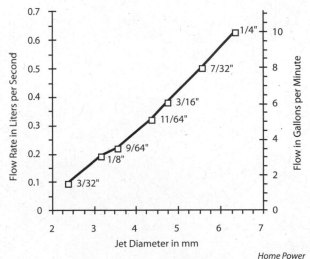

Home Power

change the jet (as the water flow changes) or to clean it. Bob-O kindly provided me with a collection of brass jets of different sizes for our trials.

We Love Lil Otto

Unlike equivalent Australian microhydro systems that I have looked at, Lil Otto is cheap (A$600), small, relatively efficient and uncomplicated. What Lil Otto clearly demonstrates is that using seasonal microhydro resources, even in Australia, should not be sneered at. This is especially true in high conservation areas such as Daintree World Heritage area where a viable alternative to grid power must be found quickly if "the sun ain't gotta shine!"[1]

Editor's Note: This article first appeared in *Home Power* Issue #52 (April–May 1996) and is reprinted with permission of the author.

Water Power in the Andes: Yesterday's Solution for Today's Needs

RON DAVIS

Going to work these days is always a bit of a thrill for me—often more than I care for. It means crossing a 15,000 foot pass over the Bolivian Andes and snaking down a muddy one lane road carved into the face of immense cliffs. "The Most Dangerous Road in the World" was the title of an old *National Geographic* article about this spectacular route.

World's Biggest Solar Machine

Actually I'm entering the world's biggest solar energy machine—the Amazon basin. Towering glacier-topped 20,000 foot peaks are clearly visible from our tropical water power demonstration site. The eastern face of the Andes so thoroughly captures the Amazon moisture that the western side—the Atacama desert—is said to be the driest place in the world. Sometimes rain only falls there a few times during an entire lifetime.

But on this side, it's just the opposite. Uncounted streams and waterfalls abound, some falling hundreds of feet directly onto the roadway! About 80 people die each year on this short section of road, since it is very narrow and slippery. Vehicles that slip off the road can simply disappear into dense vegetation 1,000 feet below. It's incredible to think that this is the only road into a tropical part of Bolivia the size of Texas.

Leaving the narrow road, it's a relief to arrive in the lovely town of Coroico at 5,500 feet, near our demonstration site. Green hillsides are covered with coffee, citrus and bananas. This also happens to be the home of Bolivia's traditional coca leaf production, so the area is much affected by the US "war on drugs."

Turgo Runner and Nozzles

Ron Davis

Campo Nuevo — Meeting People's Needs

Over 15 years ago, Diane Bellomy and I founded Campo Nuevo. We started our family-sized appropriate technology organization to improve lives by bringing simple technology to Bolivia's indigenous people. We teach them how to use their local natural resources for energy. We want to show people how easy it is to employ the abundant small local sources of water power to improve their lives. This can help make it possible for them to remain on their land and in their own communities.

We are working with Aymara-speaking native Americans, one of the largest and most intact indigenous cultures in the Western Hemisphere. Notable for having withstood the Incan conquest and later the Spaniards, the Aymaras are now succumbing to the pressures of modern global economics. Like rural people all over the developing world, they are being forced to relocate simply to survive. They usually migrate to a desolate 13,000 foot suburb of La Paz in order to compete for unskilled, low paying and often temporary jobs.

A New-Old Solution

Although they may not realize it, what visitors to our demonstration site see is not really new. It's actually a revival of the nearly forgotten traditional use of water power. For thousands of years before the invention of centrally generated electricity, water power was employed to directly run machines, something it does very well.

What is new is the development of a modern, low cost turbine specifically for this purpose — a *motor* driven by water power. We call it the Watermotor. It can provide the energy to drive a variety of machines, replacing the mid-sized electric motors upon which nearly all modern production depends.

Lester Pelton, who invented the Pelton wheel, produced a variety of these water-powered motors. They were in use before 1900, powering individual machines. Pelton even used one to run a sewing machine! The direct drive hydro units were replaced by electric motors after centrally produced electricity became the norm.

Few people realize how closely rural poverty is related to the lack of machines necessary for local production and services. In the developing world, the power grid is usually confined to cities and large towns. Rural people still use muscle power as everyone did in the past, and they do without electric lights.

Cutaway View of the Watermotor

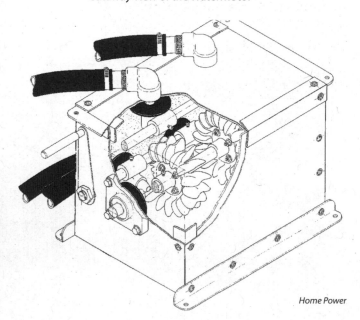

Home Power

The need to generate cash to buy anything they don't produce themselves causes a focus on cash crops. This further reduces their self-sufficiency, encouraging a downward spiral towards dependency on a system that cannot be depended upon!

Demo Site

Water power is nature's most concentrated form of solar energy, and by far the easiest to convert into usable mechanical power. At our new Campo Nuevo demonstration site, we are featuring practical machines, directly powered by water. There are woodworking tools, air compressors and water-powered water pumps. We also run an auto alternator to charge batteries and provide lighting. This can be switched on when mechanical power is not being used and is driven by the same belt drive that powers the tools.

The main attraction at our site is a Watermotor driving a small multipurpose woodworking unit. The machine is suitable for producing doors, window frames and furniture — necessities usually purchased from the city. It processes locally grown timber instead of wood carried up from the Amazon forest.

The Watermotor at our demonstration site is provided with power from a water source located 65 feet above the machine by four 170 foot long 1½ inch polyethylene pipes. At the heart of our turbine are two Energy Systems and Design plastic mini-Pelton wheels, mounted on a single shaft and driven by two water jets each. With a flow of 82 gallons per minute, we get power similar to a ¾ horsepower electric motor at about 1,450 revolutions per minute. Unlike an electric motor, the Watermotor costs nothing to operate and can't be burned out by hard use.

It's Not Easy

Not much of this area is served by roads or the power grid. The US owned (and US priced) power generating system has little incentive to provide long distance lines to a widely scattered and typically impoverished rural population. Water power is the sole available practical source of energy to run machines. There is not a good wind resource in the mountain valleys, and PV is just not economical compared to the abundant water power here.

There are major obstacles to the introduction of unfamiliar technology to an indigenous population that has traditionally used no machines of any kind. These people have little money to invest in anything that does not promise a practical return. In addition, the Aymaras are unlikely to be reached by advertising in the city newspapers. This is why we felt that a local demonstration site was necessary.

Other problems are encountered when machines, however useful, need to be professionally installed, maintained or repaired. Outside the city, such services are frequently unreliable, hard to come by and expensive when available.

Keep It Simple

In order to overcome these obstacles, we designed the Watermotor to be user installed, maintained and repaired. Because it is locally produced from common materials, all parts can be easily replaced. Only the Pelton wheels need to be obtained from other than local sources. A Watermotor can be made with hand tools and a drill press, though some welding is required. Most builders will find it convenient to have the hubs which connect the Pelton wheels to the shaft made by a local machine shop.

The efficiency of direct drive water power is a big advantage. A surprisingly small amount of water falling a short distance can produce the 0.5 to 3.5 horsepower of mechanical power required by most common machines. This means that many potential water power sites are available, and a minimum of civil engineering is required. Water is carried to the turbine by low cost, easily transportable plastic pipes. Rigid large diameter penstocks which require skilled installation are not necessary.

Other projects by Campo Nuevo include ferro-cement water storage tanks, ram pumps, hand powered water pumps, electric and treadle spinning machines and adobe brick and plastic greenhouses.

The Watermotor itself is very simple to build, operate and maintain. It functions efficiently in a variety of water power situations. By merely experimenting with easily changed water jets of different sizes, it is possible to vary maximum power output. This also allows the turbine to maintain efficient output over seasonal water flow variations. Control handles connected to the jets are used to divert water flow away from the Pelton wheels, cutting power.

Power Output

Regarding output and efficiency, you can determine how much energy you could get from a particular water power source by using this formula:

Horsepower equals total head in feet times flow in gallons per minute times efficiency in percent times 0.18 divided by 746

Several things need to be considered along with this formula. Pelton wheels are usually about 75% efficient. There will always be some pressure loss due to friction in the water supply pipes. Your local supplier should be able to help calculate this for different products. Tables for pressure loss in pipes of various sizes can also be found in alternative energy catalogues.

The power output of the Watermotor depends on the fall and the amount of water used to run it. Here are some examples of other possible installations and the energy output that they would produce:

- A 100 foot fall and 110 gallons per minute would produce 2 horsepower at 2,050 revolutions per minute.
- A 150 foot fall and 184 gallons per minute would produce 5 horsepower at 2,550 revolutions per minute.

The Basics

The Watermotor can be used to drive most stationary machines normally driven by an externally mounted electric motor or small gasoline engine in the 0.5 to 3.5 horsepower range. Power output can also be increased by simply lengthening the housing to accommodate more Pelton wheels, without basic design change.

Machines are driven by standard belts and mounted directly on or beside the turbine housing. The shaft between the Watermotor and the tool is ⅞ inch, and the housing is about 12 by 12 by 14 inches. The turbine must be mounted to accommodate the outflow without having water back up. We use a cement box as a tailrace, with a four-inch drainpipe which returns the water to the stream.

Make the Comparison

How does the Watermotor stack up against the competition? I asked a couple of RE experts to give me the rough cost of a wind or PV system capable of producing 2½ horsepower of mechanical energy 24 hours a day, including installation in rural Bolivia and technical expertise for maintenance and repair.

Richard Perez of *Home Power* said, "Well, the PVs alone will cost about US$35,000. And the requirement for 24 hour power at that level means a very large battery bank which will bring the system cost up to around US$70,000. And we still need to add small stuff like racks, inverter and controls. Overall, I'd say about US$80,000. It really points out how cheap hydro is."

North American wind power guru Mick Sagrillo said, "My guess, using off the shelf equipment, would be that you'd need a ten-kilowatt Bergey Excel. While it's larger than what's needed, it's cheaper than putting up several smaller turbines. The cost for both genny and controls is about US$20,000, less tower, wiring, batteries and balance of systems components. Total system cost would be roughly US$35,000. The one message I always deliver at my wind power workshops is that if anyone has a good hydro site, they're in the wrong workshop. While wind is cheaper than PV, it's no comparison for a hydro site with a 100% capacity factor."

Now this is not a scientific comparison, and these are admittedly rough figures. But the Watermotor can produce 2½ horsepower continuously — with a system cost of less than US$2,000. It's user installable, maintainable (two lubrication points) and easily repairable. It has only one moving part, can be locally produced in a small shop and is immune to damage from hard use. Also consider that PV and wind equipment are imported, and that there's a good chance of damage from misuse or poor maintenance.

Watermotor-type designs were abandoned about 100 years ago in the developed world in favor of electric motors. To the best of my knowledge, there are no machines equivalent to the Watermotor being produced today. Generally, very few products, no matter how useful, are produced with the aim of promoting self-sufficiency among the world's rural poor.

Water Power to the People

The best advertisement for our water driven machines is for them to be seen hard at work by the many people passing the demo site daily. Woodworking machines in particular have a substantial per-hour cash value. Because the Watermotor is immune to damage from hard use, it is suitable to rent or lease. At current rates, the entire cost of a Watermotor installation should be recovered in only a few months.

We expect visitors to our demonstration site to have their own ideas about how they can use the Watermotor. The experience gained at this site will provide us with knowledge and incentive to build similar sites in other parts of Bolivia. Plans are available — contact us for more information about building and using the Watermotor.[1]

While Bolivia is especially rich in water power resources, many other parts of the world have similar conditions, and similar needs. We would like to see this clean, self-renewing, easy to use natural resource made available to all.

Editor's Note: This article first appeared in *Home Power* Issue #71 (June–July 1999) and is reprinted with permission of the author.

13

Zen and the Art of Sunshine

PHILLIP SQUIRE

Zen Mountain Center is a Buddhist retreat and training centre located in the San Jacinto Mountains of southern California. We are nestled in a steep-walled canyon with meadows, chaparral and dense stands of pine, cedar and oak. The canyon is 1.1 miles from the nearest utility hookup.

The property, covering 160 acres of pristine wilderness, was originally purchased in 1981 by the Zen Center of Los Angeles as a summer mountain retreat. It has now grown to include a year-round population of about 15 monks and lay residents and up to 50 visitors and residents during retreats and workshops. Early in the development of the property, we conducted an environmental impact study to establish eco-friendly limits on the number of people and buildings to be supported by the canyon. This was to ensure that the awe-inspiring beauty of this land and the health and well-being of its furry denizens would be preserved.

Electricity for the New Center

Few people, even back in 1981, were willing to live without electricity. With these limits in mind, we started to consider how to power all the functions of a modern community. Our options included:

- Bringing in utility electricity from the road end by either utility pole or buried cable
- An engine/generator, battery and inverter combination
- PVs, batteries, inverter and an engine/ generator

Hydro and wind power were not deemed feasible, since both wind and water flows in the canyon are fairly intermittent.

The upfront cost for extending the utility grid up the road in 1982 was at least US$66,000 and is approximately double that today. PV was the second most expensive option at US$15,000. The engine/generator, battery, inverter option turned out to be the least expensive in the short term, but the prospect of noise and air pollution in the quiet of the canyon ruled this out as a long-term solution. However, engine/generators were used in the very early days as an initial power source.

The PV system was eventually selected as the most viable option due to lower initial cost compared to bringing in the utility line. The use of PV would also require significant energy conservation and ensure that we would not just be transplanting suburbia into a mountain environment.

Heating and Cooking

Following the decision to go with solar electricity, the obvious choice for heating, cooking and refrigeration was propane. In the early days when winter residents were few, woodstoves were used for heating. But with few exceptions, gas heaters have been phased in over the years. The first reason for this is that we are in a very high fire danger area, with only one road out down the canyon. A second reason is smoke, and a third is the logistics of gathering, cutting and distributing all that firewood.

We use propane for water heating, to fire one commercial and two domestic stoves and to run six assorted gas refrigerators and one freezer. We have two 500 gallon tanks that are filled on a regular basis by a supplier from our local town of Idyllwild ten miles away. These feed underground gas lines to serve accessible buildings, with bottled propane providing for outlying cabins and trailers.

Developments in Electricity Generation

At present, Zen Mountain Center has about 20 separate structures, including meditation halls, kitchen and dining facilities, guest accommodations, a workshop, meeting rooms and a large bathhouse. All these structures were built by our Abbot, Charles Tenshin Fletcher, a building contractor by trade, and 20 years of hardworking Zen Center members. Most of these buildings are located on about an acre in the lower end of the property and are powered by what we call the *main system*.

The remaining structures (two cabins and two trailers) are located in the upper part of the grounds. These are powered by their own individual PV arrays due to their distance (up to 1,500 feet) from the main electrical system. These systems consist of two or three Arco M-75 panels, a Sun Selector M16V 3.0 charge controller and two six-volt Trojan T-105 batteries (220 ampere-hours at 12 volts).

These smaller systems have given very little trouble since they were installed. The only maintenance that has been needed is replacement of one faulty charge controller, adding water to the batteries and very occasional charging (by portable engine/generator) during periods of low sun and heavy usage.

The main system has gone through a number of changes and expansions since 1981, but originally grew from a consolidation of several different PV systems dotted around the lower property. In 1993, when it was first put in place, it consisted of a combination of 18 Arco M-75s and eight Arco1 6-2000s on a fixed mount in our lower meadow—located 100 feet from the batteries and connected with four, #2 USE direct burial cables. About half of these modules were used and had considerably reduced outputs from their rated values.

A 30 amp charge controller and voltage monitor (Speciality Concepts SC2) and a 3,000 watt Westec Systems W3000-24 120/240 inverter were the electronic brains of the system. The system also included eight, six-volt Trojan L-16s (700 amp-hours at 24 volts), a 6.5 kilowatt Honda ES6500 gasoline engine/generator as our backup, a 24 volt battery charger and a Todd Engineering power switch to change the output from inverter to generator (or vice versa) during charging. The output from this system was delivered as 240 VAC to all individual building breaker boxes, and powers 120 VAC loads inside the buildings.

This system provided learning experiences for PV novices and delivered most of the electricity in the canyon. It was totally manual, and required monitoring to ensure that the batteries did not get too discharged during low sunshine days.

Over the first two years the batteries were in use, they lost most of their useful capacity. The main problems appeared to spring from a faulty battery charger coupled with not enough knowledge about how and when to recharge the bank. Consequently, the bank never received a full generator charge, and certainly never an equalizing charge. As usage increased, the PV array became increasingly undersized, which meant that there was never enough solar input to bring the batteries to a full state of charge. Ruining a battery bank in two years was quite an expensive lesson.

One other problem was that there was a short loss of power during the generator/inverter switchover, which meant keeping the computer people informed about the upcoming reboot. Also, if heavy power tools or the washing machines were being used, the lights would flicker and the computers would go down. We found that the modified square wave output from our inverter was not compatible with separate, stand-alone computer UPS systems.

Upgrading the System

In the summer of 1995, we decided to bite the bullet and go high tech. We calculated that in the next five years, we could be using anywhere from four to eight kilowatt-hours a day,

Zen Mountain Center Loads	
Location Average watt-hours per day	
Workshop	
Circular Saw	71.4
Band Saw	71.4
Compressor	71.4
Drill	53.6
Lights	28.6
Generator shed	
Inverter idle load	384.0
Wash house	
Washing machines	857.1
Office	
Computer	1,650.0
Printer/fax	157.1
Lights	114.3
Phone	96.0
Answering machine	48.0
Kitchen	
Lights	228.6
Vent fan	214.3
Blender	128.6
Mixer	71.4
Bathhouse	
Lights	320.0
Community room	
Computer	214.3
Lights	137.1
TV/VCR	102.9
Dormitories	
Lights	411.4
Staff Housing	
Lights	900.0
Vacuum cleaner	600.0
Computers	457.1
Coffee maker	400.0
Answering machines	288.0
TV/VCRs	114.3
Stereos	75.0
Meditation hall (large)	
Lights	731.4
Meditation hall (small)	
Lights	68.6
Total average watt-hours per day	**9,066**

Phillip Squire

depending on the number of residents. The maximum power draw of around three kilowatts would occur if several power tools and the kitchen's vent fan were running at once. So we needed an inverter that could provide that sort of output over a short period of time, and enough surge to get the motors started.

We also wanted an automatic starting option for the generator, and as pure a sine wave as we could get. The battery storage should be enough to get us through a couple of days of cloudy weather, which meant purchasing at least 20 kilowatt-hours (approximately 800 amp-hours at 24 volts).

After researching several options and raising the necessary cash, we purchased and installed a Trace SW4024 inverter and a propane powered seven-kilowatt Kohler 7CCKM generator with an hour timer. The batteries were hooked up to the inverter through an Ananda SF400-T disconnect. The output was then fed into our main breaker box located on the outside of the workshop, and from there to the rest of the property.

The old batteries were kept in place with the hope that an equalizing charging regime would bring them back to life. When that failed, we tried the EDTA treatment, but this also had little effect. It seemed that most of the generator and solar electricity that was pumped into them just disappeared. So digging into our pockets again, we bought as many new batteries as we could afford without seriously undersizing the system. Sixteen Trojan L-16s were purchased, which gave us a total storage of 1,400 amp-hours at 24 volts.

This amounted to a useful storage of around 700 amp-hours (17 kilowatt-hours) if we were to use the recommended 50% discharge depth. We also built a new generator shed with one room for the generator, inverter and meters, and one room for the batteries. Four additional PV panels (Arco M-60s) from a system powering the kitchen and guest accommodations (which was subsequently hooked onto the main system) were added to the main PV array, giving us a maximum output of around 30 amps at 24 volts.

In October 1996, we purchased a Trace C40 charge controller with digital display and a Cruising Equipment E-meter. Together, these meters allowed monitoring of total amp-hours produced and consumed, and showed that we were using a lot more than we'd estimated.

On the average day, instead of the four to eight kilowatt-hours we'd calculated, our total energy consumption was more like eight to nine kilowatt-hours. Because we were producing only four to five kilowatt-hours on a sunny day, the center was seriously underpowered. This explained why the generator was running every two to three days. With this energy deficit in mind, we have been purchasing panels as we can afford them and installing compact fluorescents everywhere.

Over the past few years, we have added two Siemens SP75s, six BP Solar 270ULs and four BP Solar 275Fs, giving us a maximum output of around 1.2 kilowatts. With the higher current through the wires, we beefed up the connection between the array and the inverter to four, #3/0 cables. We also split the main array into two subarrays (300 watt and 900 watt), and purchased an RV Power Products Solar Boost 50 as the additional charge controller.

With the extra panels, we are now about 100% renewable during summer. We will still be slightly underpowered during winter and when we host large retreats. Our goal is to be 100% green powered on an average day during winter. We do, however, plan to keep covering the intermittent spikes in energy consumption during large guest retreats with an engine/generator rather than another subarray of solar-electric panels.

Since we added the Solar Boost 50 charge controller, I notice that there can be as much as a 15% increase in output on cold days during the winter when the batteries are low. There isn't much gain during the summer when the batteries are full most of the time and the voltage is high. But overall, this unit is worth the cost.

Energy Distribution

We are using three, #2 cables from the inverter to a main service panel. From there, using the same cable size, electricity is fed underground to subpanels at individual buildings and complexes around the property. The main panel is fused with 100 amp breakers.

The old Westec inverter used to give us 240 VAC, which we delivered using the three #2 cables with two hot legs, and one neutral. With the new 120 VAC Trace inverter and all the cable direct burial, we decided to jumper the original hot legs together at the main breaker box and at individual breaker boxes. This gives us two, hot #2 cables and one #2 neutral. Each subpanel is grounded individually.

Quirks, Glitches and Pleasant Surprises

A few other interesting changes occurred in our operating system after the upgrade that are worth mentioning. Just before the upgrade, we installed an extractor fan and swamp cooler in our kitchen to live up to the requirements of the Department of Environmental Health in Riverside County. These two fans together draw about 2,000 watts. This is a tremendous amount of energy for a small system like ours, so we use it as sparingly as possible. With the experience we've gained over the years, we'd have purchased different fans, but we rarely use them, except when things get really smoky.

With the old Westec inverter, it was impossible to start the two fans at the same time. This may have been because the high surge, undersized wiring or the modified square wave was not to the liking of the fans. This caused some inconvenience, since the motors were started by the same switch in the kitchen, and it meant climbing onto the roof and disconnecting one while the other started up. Also, if the fans were restarted after a short break in operation, there were loud groans of protest from the motors. We blew up one capacitor this way. However, with the advent of the new Trace inverter, all these problems miraculously disappeared.

The sine wave output has now allowed installation of UPS systems for the computers. UPS systems wouldn't work properly with the old inverter — something to do with heat buildup and sensitivity to the quality of incoming power. These UPS systems are an absolute necessity, because we constantly experience flickers in the power whenever a washing machine or power tool starts up. This could be caused by a slight undersizing of the underground cable to the various buildings, resulting in a voltage drop as the motorized culprit kicked in.

This does not occur when the generator is running, which puzzled us at first, since the inverter has a higher surge capacity than the generator (ten kilowatts compared to less than seven kilowatts). But we came to find out that when charging batteries from an engine/generator, Trace SW series inverters will back off on the charge rate and actually draw energy from the batteries to assist the generator in powering large loads if need be.

The inverter has overload protection and shuts off if it experiences a load of more than ten kilowatts (in other words, if there is a short in the system). This is fine except that if there is a short somewhere on the property, the inverter will shut the whole system down before an individual magnetic breaker in one of the buildings can do its job. This means that it's harder to find the source of the short. And once you know where the short is, it can be a long walk to restart the system, and it's an inconvenience to everyone else at the center.

Even the waveform of the Trace SW4024 sine wave inverter may not remove the nasty hum from a stereo amplifier. From experiences of people around here, it appears that the hum depends on the brand and model of stereo that you buy. I bought a Sony boombox, only to hear a loud hum that would not go away even when securely grounded. I exchanged it for a similar product from Aiwa, which is as quiet as a mouse using CD and radio, but still emits a slight noise running the cassette deck. It seems worthwhile to experiment, especially if you're going to spend a bit of money.

Engine/Generator Lemon

Our Kohler engine/generator spent a fair amount of time in the repair shop immediately after its purchase. We had to get a completely new one when one of the cylinders developed a crack and sent oil and smoke pouring out of the generator shed. The replacement model developed some problems too. One spark plug was continually being plugged with black carbon deposits every 30 hours of operation. This occasionally resulted in misfiring and required regular replacement. Oil was also appearing around the breather tubes, suggesting that perhaps the rings were not properly seated.

After an almost complete rebuild (under warranty), it appeared that the problem had been mostly taken care of. However, after perhaps a year of trouble-free operation, the Kohler really gave up the ghost, due to arcing (we think) in the main coils. So we purchased a Generac Primepact 55LP with a three-year warranty.

But sadly, this too has been giving us headaches — with bad thermal breakers, oil leaks and other unidentifiable problems. So we're hunting around again for something that really works — any suggestions? All the more reason to keep on working towards 100% renewable energy.

Wind in Our Future?

Zen Mountain Center is committed to reducing our reliance on the engine/generator, so we are increasing the size of our PV array as the money becomes available. Recently we've decided to look again at feasibility of wind and water as energy sources. As mentioned earlier, we didn't think we had enough of either resource for useful energy generation. However, we started to experiment with wind, using an Air 303, just to be sure. In the cooler months, there are some terrific gales (Santa Anas) which come from the high desert and sweep over our mountains and down into the canyon.

In January of 1999, we installed an Air 303 in the top of one of the tallest pine trees near the electrical shed. Since the Air is priced about the same as a single solar-electric panel and has the potential for putting out 400 watts during very strong winds, it appeared worth the initial investment.

The downside to this wind generator is that it is not at all quiet. It has a distinctive moan when wind speeds get above 20 miles per hour, which has been a little irritating to those used to the whisper and rush of wind in the pines.

The Air 303 is hooked up via 100 feet of #8 UV Romex running down the tree (stapled as required by code) through a removable, three-point plug and socket (for quick disconnect from the electrical shed in case of pending lightning) and a 20 amp fuse to the battery bank. We also installed an Ananda LA100V lightning arrestor in line, in case we forget to disconnect the wiring during storms.

We recorded the output of the Air 303 using an E-Meter for the first year. We received about 1,500 amp-hours (at 24 volts), or 36 kilowatt-hours for a 12 month period. In the same time period, our PV array gave us 1,632 kilowatt-hours. Investing the same amount of money that we spent for the Air 303 in another solar module would yield perhaps three times the annual energy output. Considering this, plus the higher maintenance requirements for the wind generator and the intermittent nature of our wind resource, we decided that solar electricity is a much better bet than wind for our site.

Photovoltaic system with hydro rock in the foreground

Jim Lakey

Microhydro

Water power is something we're now looking into. The wells from which we gravity feed potable water for the center (horizontally drilled into the granite at the back of the canyon) are at an elevation about 180 feet higher than the electrical shed. And we already have a two-inch pipe for water supply and fire protection running around the property.

We gave Don Harris of Harris Hydroelectric a call and discussed the feasibility of installing one of his microturbines. Don came down and talked us through the installation, and we have now connected one of these amazing little pieces of equipment to the water system and the electricity system.

Don gave us several different nozzle configurations for our system, based on our available head and flow rates. We are still in the testing stages, but will probably be running a flow rate of six to nine gallons per minute for an output of 100 to 150 watts. This could provide a major portion of the electricity we need in one fell swoop!

The microhydro turbine is directly wired to the battery bank via 100 feet of #2 cable through a 15 amp breaker. Since the microhydro must be connected to a load at all times to control the turbine's revolutions per minute, we also installed a second Trace C40 (in diversion mode) and an Enermaxer controller with a 1,440 watt diversion load, an air heater and fan. Any excess electricity produced by the turbine is dumped to this load once the batteries are full.

The bulk and equalize set points on the diversion Trace C40 are set a couple of points above the set points on the solar charge controllers and Trace inverter. This ensures that the output from the solar array or generator does not start discharging into the heater/fan diversion load when battery voltage approaches bulk or equalize voltage. When we run an equalizing charge with the propane generator (we don't have enough capacity to

do this with renewables) we make sure to hit the equalize button on the diversion Trace C40.

The success of this project, however, depends on whether we have enough water in our aquifer. Our wells are our only water supply, and this is southern California. We are not sure what kind of effects moving this amount of water from one end of the property to the other will have on the riparian ecosystem around the wells and on the viability of our water supply.

We're planning to only operate the hydro system during the winter, once the rains and

Close-up of turbine installation

Jim Lakey

Zen Mountain Center Main System Costs	
18 Arco M-75 modules	$5,400
Trace SW4024 inverter	$3,500
Generac Primepact 55LP propane generator	$2,800
8 Arco 16-2000 modules	$2,400
16 Trojan L-16, lead-acid batteries	$2,400
6 BP Solar 270UL modules	$2,100
Boxes, breakers, fuses, conduit, cables	$1,500
4 BP Solar 275F modules	$1,400
4 Arco M-60 modules	$1,200
Harris microhydro turbine	$800
2 Siemens SP75 modules	$700
SWWP Air 303 wind generator	$500
2 Trace C40 charge controllers	$370
RV Power Products Solar Boost 50 controller	$350
Ananda APT SF400T disconnect	$300
E-Meter amp-hour meter	$190
Enermaxer diversion controller, 1,440 watts	$150
Ananda LA100V lightning protector	$60
Total	**$26,120**

Note: Cost (in US dollars)

Phillip Squire

Zen Mountain Center Renewable Energy System

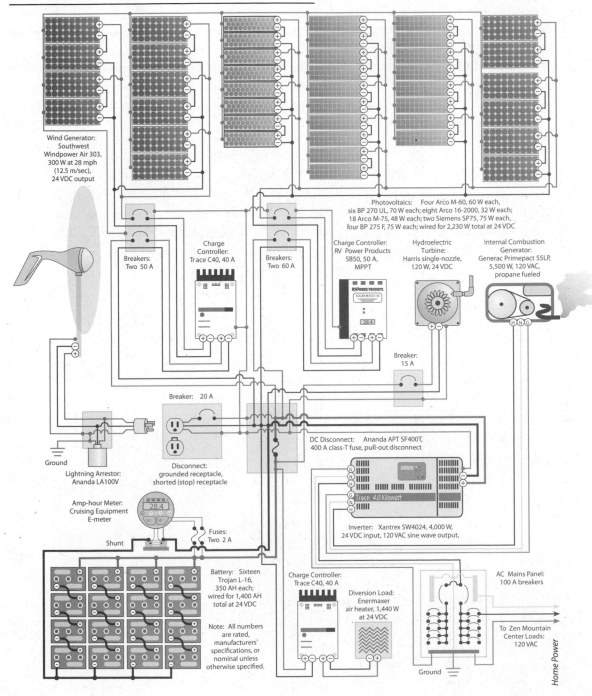

Wind Generator: Southwest Windpower Air 303, 300 W at 28 mph (12.5 m/sec), 24 VDC output

Photovoltaics: Four Arco M-60, 60 W each, six BP 270 UL, 70 W each; eight Arco 16-2000, 32 W each; 18 Arco M-75, 48 W each; two Siemens SP75, 75 W each, four BP 275 F, 75 W each; wired for 2,230 W total at 24 VDC

Charge Controller: Trace C40, 40 A

Breakers: Two 50 A

Breakers: Two 60 A

Charge Controller: RV Power Products SB50, 50 A, MPPT

Hydroelectric Turbine: Harris single-nozzle, 120 W, 24 VDC

Internal Combustion Generator: Generac Primepact 55LP, 5,500 W, 120 VAC, propane fueled

Breaker: 15 A

Breaker: 20 A

Ground

Lightning Arrestor: Ananda LA100V

Disconnect: grounded receptacle, shorted (stop) receptacle

DC Disconnect: Ananda APT SF400T, 400 A class-T fuse, pull-out disconnect

Amp-hour Meter: Cruising Equipment E-meter

Shunt

Fuses: Two 2 A

Inverter: Xantrex SW4024, 4,000 W, 24 VDC input, 120 VAC sine wave output,

Battery: Sixteen Trojan L-16, 350 AH each; wired for 1,400 AH total at 24 VDC

Note: All numbers are rated, manufacturers' specifications, or nominal unless otherwise specified.

Charge Controller: Trace C40, 40 A

Diversion Load: Enermaxer air heater, 1,440 W at 24 VDC

AC Mains Panel: 100 A breakers

To Zen Mountain Center Loads: 120 VAC

Ground

Home Power

snows arrive. We'll keep a close eye on biological health, well flow rates and surface water levels in the stream bordering the well site. It may be that we won't feel comfortable with running the turbine even during the winter for any length of time, but at the very least it will be a backup supply for when the wind isn't blowing, the sun isn't shining and the generator has decided to quit.

Renewable Community

Living in a community with renewable energy is an ongoing education in how to balance a large and diverse group of needs with the realities of a finite amount of energy. Informing visitors and residents regularly about the need for conservation, and the necessity for using energy efficient lighting and appliances are parts of a process that keeps us aware that electricity comes at a cost.

At Zen Mountain Center, we are directly aware of that cost, since we produce the stuff. Everyone who visits here experiences living with renewable energy. And if we do nothing else than alert people to the fact that every time we flick a switch, we are having an effect on the environment, it will be well worth it.[1]

Editor's note: This article first appeared in *Home Power* Issue #92 (December 2002–January 2003) and is reprinted with permission of the author.

14

Power to the People

DON HARRIS

In February 1987 we had the opportunity to plan and install a small DC hydroelectric generator on a rural dairy cooperative in Nicaragua. The ranch had belonged to a Minister in the Somosa Regime. After the Revolution, the land was distributed to the workers who had formerly lived under conditions resembling serfdom. The cooperative has a total of 34 families, nine of which presently live on-site. Our objective was to provide enough user friendly electricity for lights and improvements for present and future families.

The Site

Eight houses are spaced about 75 feet apart in two rows of four each. The ninth house is over ¼ mile away and was the original hacienda. It also serves as a gathering place for meals and fiestas. A creek runs within 500 feet of the nearest house and a 3,000 foot long, nearly level flume passes between them. The flume was built to feed the swimming pool. It now also provides agricultural water for the dairy operations.

The Project

In November, 1986 I was contacted by some friends who were planning the *Power to the People* project and needed information and hardware. The project was sponsored by Technica, a Berkeley, California-based technical assistance organization. They expedited all the complexities of getting to and from Nicaragua. Always one to travel, I joined. Kate, a project organizer, had been at the site previously on a house construction project. From her memory, we had enough site data to build the turbine and collect other necessary parts.

A prime design consideration was to make the system locally serviceable. This precluded the use of some of the fancier electronic equipment that is so useful to us in the USA. The exception to this was the Enermax charge controller. We had to control battery overcharge, and these units are nearly indestructible.

Because of the US embargo, Delco alternators, which we usually use, are very scarce. Japan trades extensively with Nicaragua, and the 40 amp Toyota alternator has the proper characteristics. We committed to 12 volt operation because of the universally available automotive lightbulbs, radios, batteries and so on. We chose edison base, 12 volt, 25 watt lightbulbs because of their reliability, but took along adaptors to convert to automotive type bayonet base bulbs, just in case.

On February 3, four days before I was to board the plane to Managua, I got a frantic message from Kate and Bill. They had driven down through Mexico and Central America the month before. They had discovered that pipe availability was a problem, and we probably couldn't get the 100 feet of head that we had planned on — perhaps as little as 20! A quick conversion back to a rewound Delco alternator produced a system that would oper-ate from 10 feet of head up and could use one to four nozzles.

My plane tickets gave me two weeks in Nicaragua. We had to plan, scavenge parts, transport everything 100 miles, install, trouble-shoot and get back to Managua in that very short period of time. Upon touching down in Managua, I was met by Bill, Kate, and Ben Linder, the first US citizen to die at the hand of the Contras. We spent the next two days in and around Managua rounding up pipe and hiring a truck to haul the 3,000 feet of four-inch PVC we managed to obtain. On the third day we headed north to Esteli and the project site.

The original plan was to run a pipe parallel to about 1,500 feet of the old flume and then pick up as much head as possible in the creek bed. We set out surveying and found that with our 3,000 feet of pipe we could get almost 100 feet of drop, our original estimate. But we also noted the rugged, almost vertical canyon walls in the gorge and the fact that we had only nine days left to get it all done. We had the full-time help of two local people and the whole community at crucial times.

Chris, a US citizen working in Central America, and Ben arrived about this time, and an alternative plan emerged. If the flume delivered far more water than the needs of the ranch, we could divert some of the water some of the time thru a 300 foot long pipe back into the creek. This would be much quicker (and thus more likely to be finished) and would save 2,700 feet of very precious pipe for other use. The flume had a diversion gate at about the right place. With a little brick work and some screen as a filter it could be used. Chris and Ben consulted the ranch elders and deter-

System Layout

8 houses with central battery shed

original pipe plan

Power Line

Hydro input

PIPE   Turbine

FLUME

CREEK

9th House

POND

Home Power

mined that they could afford 12 hours a day operation in the dry season. We quickly surveyed and found we had 78 feet of head. We had a system!

The practical (50% efficient) potential from 300 feet of four-inch pipe and 78 feet gross head is about 2,100 watts. This would be using 450 gallons per minute at 54 feet net head. Our unit using four nozzles can use up to 160 gallons per minute and could, with the right alternator and at the right voltage, produce 800 watts. Our commitment to 12 volt operation and our use of the ultra low head alternator limited us to eight amps. This latter limit is due to the small diameter, long wire in the special stator winding.

We had to go 500 feet from the turbine to the batteries and up to 250 feet from the batteries to the houses, a long way for 12 volt transmission. We had 3,000 feet of 12-2 Romex wire which translates into 9,000 feet of #12 wire including using the ground wire as a conductor. We did get 200 feet of #10 single strand wire in town, but it is scarce and it seemed almost antisocial to use too much. After playing with the numbers, the best choice seemed to be one run of Romex to each house, two conductors + and one —. This is about .7 ohm resistance in the worst case. The 25 watt lights we used are six ohms, so wire losses are a little over 10%. Though not ideal, this was acceptable. The practical result is slightly dimmer lights that will probably last longer because they are running at 12.5 volts. The two 100 amp-hour gel cell batteries are held at 13.8 volts by the Enermax regulator.

The remaining wire provides four runs of Romex from the turbine to the batteries – six conductors +, six conductors —. This is about .26 ohms. With eight amps output the alternator runs at 16 volts to deliver 13.8 to the batteries, about a 14% line loss. Again not ideal, but acceptable in this case.

The pipe runs almost level for 240 feet, gaining perhaps 20 feet of head and then plunges almost vertically for 60 feet into the creek gorge. A very steep switchback trail goes part way down the canyon, but the last 20 feet are so steep that we had to build a ladder to even see if there was a spot to mount the turbine. The wood was milled on-site, freehand with a chainsaw. I wish the wood I buy at the lumberyard were all as straight. Fortunately there was a convenient little flat at a spot about 20 feet above the creek. No one remembered seeing the water that high in the wet season.

We had to tie the pipe to trees to support the weight of the long vertical section and build a sturdy shed roof over the unit because our working resulted in a continual avalanche on the site. Indeed, someone often watched as others worked to warn that boulders were on the way!

Kate worked on building the light and switch wiring for the houses. She surveyed each family for their choice of light placement. Because of the mild climate, most people lived more outside than in the house. Someone came up with the ingenious idea of knocking out a high wall board, allowing light both inside and out, and everyone followed.

As our Romex was not direct burial rated, we encased it by dragging it through one-inch plastic pipe for protection before burial. This was a most strenuous operation, especially the four wire section from the turbine to the batteries. Each house was individually fused on the + side at the battery end, and a protective box was built around the storage/ distribution complex. This protected the hardware from the pigs which will aggressively explore anything they can get at.

Finally, one day before we had to leave, the moment came, we turned the valve and the turbine gurgled and belched its way up to eight amps in a few minutes. We were on line! Later that day we connected the houses with nary a glitch, and the neighborhood lit up! We had forgotten the ninth, more distant house until the last week. Bill located a battery in town and we set up a shuttle system to the charging station. The following month, Dave Katz of Alternative Energy Engineering went down with solar panels and—but that's another story.

The final statistics are 125 watts at the turbine using 21 gallons per minute and 77 feet net head. This rather low 40% efficiency is due to high losses in the special wound, low head stator. One hundred and ten watts are getting to the batteries after wire losses. The system operating 12 hours will produce 1.3 kilowatt-hours a day, enough to allow each house six hours of light. This far exceeds the perceived needs of the families. The last day at the ranch was a festive occasion in celebration of the project. We left for Managua with warm feelings and happy memories of this time with our Nicaraguan friends.

What It Cost

If translated into North American terms, the total hardware cost of the system was US$2,850. It breaks down something like this: the cost per house is US$316 including delivered power, house wiring and one set of spare lightbulbs and fuses.

Maintenance costs should be primarily battery replacement every five to seven years, plus occasional light, fuse and alternator part repairs. The leaves need to be cleaned off the screen periodically and possibly a nozzle unplugged. Time will tell.

Some Final Thoughts

One late night about a week into the project we were awakened by two earth shaking explosions. The next day we found that the Contras had blown the main power lines 15 miles from where we slept. These were no firecrackers. Much of Northern Nicaragua was down. When we left for Managua a week later, the only evidence of electricity I saw was at our project. A striking impression was that of hundreds of people hauling drinking water on their backs for miles. The city's water treat-

System Costs

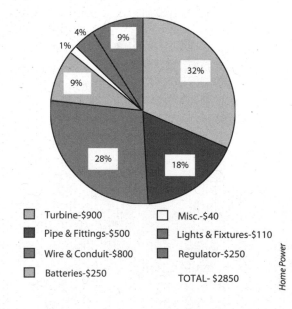

Turbine-$900
Pipe & Fittings-$500
Wire & Conduit-$800
Batteries-$250
Misc.-$40
Lights & Fixtures-$110
Regulator-$250
TOTAL- $2850

Home Power

ment plant is electrically operated. Two facts were evident:

- The real burden of terrorism is borne by the common people.
- Those of us that produce our own power are free indeed in times of civil strife.

Ben Linder was at the site for two days in the early part of the project. We sat one night and talked about the world. He shared a profound understanding of the situation in Central America. He wanted so much to heal the wounds. We made plans to apply water power to grinding corn and coffee. Ben brought lights and happiness to the people, and they loved him. Not only did he electrify several villages, but he helped bring the Children's Circus to Nicaragua. He was the best kind of ambassador the US could possibly have. He is missed there as well as here.[1]

Editor's note: This article first appeared in *Home Power* Issue #8 (December 1988–January 1989) and is reprinted with permission of the author.

A Batteryless Utility
Intertie Microhydro System

KURT JOHNSON, WITH PAUL HOOVER

Kitty Couch lives near the end of a gravel road outside Burnsville, North Carolina. Plum Branch drops out of a ridge facing the Black Mountains along the east boundary of her property. It is a small, seasonal stream with a flow range of around 60 to 130 gallons per minute. At her location, a 75 foot head over a distance of 1,000 feet is available.

Kitty is a potter and uses electric kilns. Between her home and shop, she had been using about 650 kilowatt-hours per month. She wanted to reduce her US$80 monthly electric bill and wanted to take advantage of the water resource of Plum Branch. She called The Solar Guys, a local renewable energy business that I run, and asked me to design and build a microhydro system for her.

Site Analysis

The site survey started by determining Kitty's property boundaries and where the stream lay within these boundaries. She had just acquired another 3.5 acres at the upper part of her property, which increased the length of the stream she owned. I took advantage of the fact that the surveyor was coming to map her

new boundaries; I met him and got him to show me the highest point of the creek on her property.

From a point within a few feet of the boundary, there was a small waterfall that would work perfectly for the Aqua Shear intake I was hoping to use. After determining the highest possible point on the stream, I needed to

Sophia likes the Aqua Shear intake — from here, 1,000 feet of pipe runs to the turbine.

Kurt Johnson

determine the lowest. The most obvious place was the existing pond which had been constructed at the bottom end of her property. It already had a four-inch PVC penstock buried along the stream for about 400 feet.

This looked like the best place to put the turbine house. All I had to do was tie into the pipe flowing into the pond and let the tailwater spill into the pond. Next I used a transit and worked my way up the hillside to the small waterfall. I determined that Kitty had 75 feet of head for her hydro site. After measuring this out along the proposed path of the penstock, I concluded that adding another 600 feet to the existing 400 foot penstock would do the job.

Hydro

Next I needed to determine the flow of the stream. By setting up a small weir to determine the volume and using a bobber and a stopwatch to determine the speed of the stream, I came up with fairly consistent figures. The stream was yielding 130 gallons per minute during the wet season.

The neighbor above Kitty's property also has a hydro system on the same stream. Since he is farther up the stream, he has less flow. But he was able to let me know that the stream usually cuts to half its wet season flow during the dry season. From this I figured that the stream would reduce to roughly 60 gallons per minute during the dry season.

Now I knew that the head was about 75 feet, the length of the penstock 1,000 feet and the flow of the stream 60 to 130 gallons per minute. I only wanted to use half of that, so I would size the system to use 30 gallons per minute in the dry season and 65 gallons per minute in the wet season.

Water Intake and Screen

The nice little 18 inch waterfall channeled the stream into a narrow flow of about six inches in width. After reading the article in *Home Power* Issue #71 on the Aqua Shear screen, I had a feeling that this was going to work out well. The screen isn't cheap, at almost US$200 a square foot. But it's worth it to have no maintenance on the intake — especially since Kitty is in her 70s and wouldn't want to be continuously cleaning the screen.

When I received the screen, which was the smallest piece I was allowed to order (one foot by one foot), I made up a mock intake box with plywood. I had to make it so that it conformed to the rocks in the small waterfall and allowed the screen to receive the bulk of the water coming over it.

Then I had to determine where the penstock would attach and how. I also made a little overflow slit above the penstock and below the Aqua Shear screen. This was because we were only taking half the water flowing onto the screen (this was controlled by the size of the orifices at the turbine). The overflow slit helps determine when the orifices need to be changed. When water stops coming out of the overflow, it's time to reduce the size of the orifice.

Once I had perfected this, I took it to a local welder. Fortunately, he had done a lot of work similar to this for the local mining industry over the years. He made a simple ¼ inch steel plate box from my plywood mockup. It had a footprint that was roughly a square foot with a back wall 18 inches high like that of the waterfall and a front wall that was determined by the 40° pitch of the square foot Aqua Shear screen. It allowed enough room for the penstock flange to be mounted and still had room for a ¾ inch slit for overflow. I later had to put mesh over the slit to keep salamanders from crawling in. The side walls were brought up about 22 inches high to create a channel to force the water toward the screen. On one side, I had to leave a small section out to accommodate a rock.

The stainless steel Aqua Shear screen filters to a 0.02 inches particle size, and can draw 350 gallons per minute per square foot. It is self cleaning, and should require little or no maintenance.

Penstock

There was already a four-inch schedule 40 PVC pipe run from the stream to a pond where the turbine now sits on Kitty's property. This pipe had been installed to feed the man-made

pond. The intake for this pipe was some 600 feet below the intake needed for the turbine.

I ran a 600 foot section of three-inch black polyethylene from the turbine intake to the existing PVC pipe. The four-inch pipe had been buried, and I was able to tie into the top of it with the poly pipe and run that up through the woods. It ran along the stream, but far enough away to not be affected by flooding.

I was not able to bury the pipe any farther up the stream because of the rough terrain (rock and laurel thickets), so I went with poly pipe. Three-inch was sufficient for the water flow, and poly was my choice because of cost, sunlight resistance, freezing durability and

Left: A hinged roof covers the two-nozzle Harris turbine. The tailrace dumps into Kitty's pond.

Right: The air release valve between the 600 feet of three-inch pipe and 400 feet of four-inch pipe

Kurt Johnson

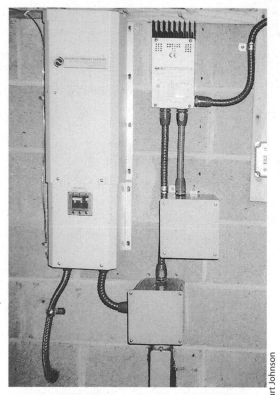

Kurt Johnson

The Advanced Energy Systems GC1000 intertie inverter and Trace C-40 used as a diversion controller

flexibility. I used 100 foot lengths of pipe with slip fittings and clamps. I secured the pipe with metal fence posts and used stranded galvanized wire with a plastic sheathing to attach the pipe to the posts. In one section, there was an old mining road. I cut a metal culvert in half lengthwise, laid it over the pipe, and buried it. This was to protect it from the frequent four-wheelers that use the road for recreation.

The only other thing noteworthy about the penstock was that where I joined the three-inch pipe to the four-inch pipe, I put a breather in so air pockets can be released manually. The outflow from the turbine is directed to the pond. The turbine is set two feet above the ground level to enable easy changing of nozzles without removing the turbine housing. The turbine house is constructed of concrete block with a hinged roof. There is very little noise, and the turbine is well protected from the elements.

Turbine and Controls

Analysis of Kitty's water resource indicated that the flow and head were adequate for a DC hydro system generating 200 to 400 watts. It would use dual 5/16 inch jets during the dry season for about 200 watts of output. During late fall and winter, up to 400 watts could be generated using dual 7/16 inch jets. The projected flow rates were 30 to 65 gallons per minute or about half the stream flow.

The system would generate five to ten kilowatt-hours per day or 150 to 300 kilowatt-hours per month. A 48 volt turbine from Harris Hydroelectric Systems was chosen. It is designed for battery charging of systems up to 48 volts and can generate up to 1,500 watts. Don Harris told me that 57 volts is the optimum voltage for this unit. In Kitty's system, the controller will not allow the voltage to go over 60 volts maximum.

Kitty needed the grid to meet her average and peak electric needs, and she was not interested in a backup power system. So she decided to go with an intertie system without batteries. This presented several opportunities and challenges. She would not need battery charging capacity in her system or a charge controller. There is no need to limit DC output voltage to that of the battery bank for this sort of system.

Using the grid as the load and in place of a battery bank also means saving the cost of batteries and having to maintain them. These factors led to choosing an inverter that did not have battery charging capability and which would accept a higher DC input voltage. This reduced the wire size needed between the turbine and the house, where the inverter was to be installed. This distance is about 225 feet, and #2 copper wire is sufficient to keep losses below 2%.

The system design is 1,800 watt, 48 volt (30 amps at 60 volts) with an air heater for a diversion load because it met the specs of the system. I felt that putting in a water heater for diversion was more trouble than it was worth because the system will rarely go down.

Smoothing It Out

Batteries in a microhydro system are like a flywheel and serve to smooth out voltage fluctuations. A controller such as the Trace C-40 can be used as a load controller and can switch to a dump load when voltage increases above a set level. In this application, the C-40 is set in charge control mode to dump power when the voltage rises above 60 volts. However, the control function is not reliable if voltage fluctuates.

When initially installed, the C-40 load controller was set to switch to the dump load at 60 volts. However, during tests it would sometimes switch even when there was no loss of power or inverter output. Worse yet, it would fail to trip when the inverter was shut down. The challenge was to smooth voltage fluctuations.

A high capacity (110,000 microfarad) electrolytic capacitor was double-lugged with the leads from the hydro turbine across the battery input terminals of the C-40. The PV input terminals on the C-40 are connected to the dump load. The C-40 now properly senses the voltage and switches to the dump load within a fraction of a second whenever the DC voltage exceeds 60 volts. In this way, it protects the turbine from overspeeding.[1]

Net Metering

North Carolina does not have net metering legislation. But the French Broad Electric Membership Corporation ("French Broad" refers to a river in the area), which serves Western counties in NC, supports distributed renewable energy systems and allows net metering. They are more than willing to let their customers sell power back to the grid. They require that customers don't produce more power than they use, have a lockable disconnect on the system accessible by the utility company and use inverters (such as Trace and AES) in their systems that are proven to not backfeed the grid in times of power outages.

They also require that you pull a permit and get an electrical inspection and an inspection by French Broad. And last but not least, they have you pay your utility bill once a year instead of once a month. So if you are out of town for a month and make more than you use, it won't confuse the meter reader into thinking that your meter turned over a full 100,000 kilowatt-hours instead of just spinning backwards a few kilowatt-hours. In the event that you do make more than you use in a year, they will charge you a US$50 processing fee. If there is still any more owed, they will write the customer a check.

I used a local electrician, Danny Honeycutt, to pull the permit and wire the AC side of the system to meet code and the grid-tie requirements. Danny also proved to be extremely valuable in helping me fine tune the running of the system. We established a great working relationship on this project, and now work together on every install I do, whether it be grid-tie, stand-alone, PV, wind or hydro. We now can also offer to wire the complete house, which the customer tends to like.

Up and Running

Kitty's system became operational at the end of May, 2000. Charles Tolley, General Manager of French Broad, came to the site to inspect and approve the system. He was pleased that the system was installed according to code and pulled the meter himself to see that the system posed no danger to his line workers during power outages.

An Advanced Energy AM 100 inverter monitor was also installed. It logs all inverter operating data. Data is averaged over 15 minute intervals and stored for up to the last 20 days. It also maintains an event log with data

on start-up and stops, grid power failures and the like. The system is generating 4.9 kilowatt-hours per day in the dry season. Earlier, at the end of the wet season, the measured output was about ten kilowatt-hours per day.

Responsible for Our Energy Use

Kitty has had an interest in doing this for many years. She loves nature and lives right in the heart of it. Over the years, she has noticed the impact that pollution has had on the local environment. This is especially evident when you go to the top of Mount Mitchell and see the damage that acid rain has done to the vegetation on the mountain.

Kitty realizes that she is a part of the reason this acid rain is here. She drives a car and uses electricity that is predominantly produced by coal (one of the main contributions to the pollution in North Carolina, which is only third to California and Texas in pollution production). Kitty wanted to do her part to fix that.

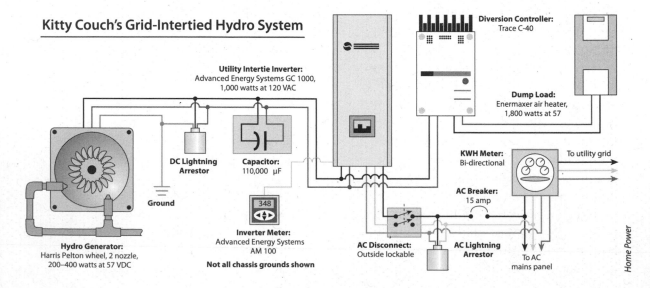

Kitty Couch's Grid-Intertied Hydro System

Utility Intertie Inverter:
Advanced Energy Systems GC 1000,
1,000 watts at 120 VAC

Diversion Controller:
Trace C-40

Dump Load:
Enermaxer air heater,
1,800 watts at 57

DC Lightning Arrestor

Ground

Capacitor:
110,000 µF

Inverter Meter:
Advanced Energy Systems
AM 100

Not all chassis grounds shown

Hydro Generator:
Harris Pelton wheel, 2 nozzle,
200–400 watts at 57 VDC

AC Disconnect:
Outside lockable

AC Lightning Arrestor

KWH Meter:
Bi-directional

AC Breaker:
15 amp

To utility grid

To AC mains panel

348

Home Power

She realized that she only had a small stream, and that her power production would be minor, but it was still important to her. Of course cost was an issue, so figuring out how to grid-tie the system without a battery bank was a key factor in making this project affordable.

The savings on her power bill have been moderate, but not insignificant. During the wet season, the turbine cuts almost US$30 off Kitty's bill (300 kilowatt-hours times 9.5 cents per kilowatt-hour), and during the dry season it is doing about half that. She now gets a kick out of looking at her meter. She can see how much power she is producing, and feels good that it is coming from a renewable source and not from burning coal.[2]

Editor's note: This article first appeared in *Home Power* Issue #80 (December 2000–January 2001) and is reprinted with permission of the author.

Couch Hydro System Costs	
AES GC 1000 inverter, GFI and disconnect	$1,785
Harris Hydroelectric turbine, 2 nozzle, 48 volts	$1,360
Penstock and pipe fittings	$1,000
Wire, conduit and miscellaneous electrical	$600
AES AM 100 inverter monitor	$540
Turbine house and penstock hardware	$500
Steel intake pieces	$200
Trace C-40	$195
Aqua Shear intake screen	$185
Enermaxer air diversion load, 48 volts	$175
Electrolytic capacitor	$56
Total	**$6,596**
Note: Costs (in US dollars)	

Kurt Johnson with Paul Hoover

16

PV/Hydro Systems and a
Visit to the Lil Otto Hydroworks

RICHARD PEREZ

I was delighted when Bob-O Schultze and Otto Eichenhofer invited us to visit some PV/Hydro systems along the Salmon River. The Salmon River runs madly through northern California, and if you want power along the river then you make your own.

The independent folks living along the Salmon have been doing just that. Brian Green, our *Home Power* photographer, and I saddled up and drove up and down the icy mountain roads to the little town of Cecilville, California. Everyone in Cecilville makes their own power. The nearest utility is over 30 miles away — through some of the most rugged mountains in the USA.

The Cecilville Scene

We met Bob-O and Kathleen at the General Store in Cecilville (population 20). Kathleen, Bob-O's wife, was doing the driving since Bob-O was recovering from an argument with a large tree that nearly cost him his leg. The Cecilville Store, hub of all neighborhood activity, is powered by a 15 kilowatt diesel engine/generator. While few folks live inside the micro village of Cecilville itself, many live up and down the serpentine one-lane road that

follows the Salmon River's course. Almost all the folks along the river have engine/generators. Many are using micro or nanohydro systems and photovoltaics.

Joyce Eichenhofer's Home

Our first stop was the home of Joyce Eichenhofer. Joyce lives right next to the Salmon. Her beautiful house is powered by homemade electricity. Joyce uses a combination of three power sources. Visible on her roof are four photovoltaic panels (two Kyocera 48 watt panels and two Solavolt 36 watt panels). This PV array produces about 168 watts, or about 12 amperes at 13.5 VDC, when under full sun. Bob-O mentioned that at Joyce's location the winter sun only shines on the panels for about 2 hours daily (summer performance is much better). During the winter, Joyce falls back on her Lil Otto hydroelectric system for power. Lil Otto turbines are made by her son, Otto, so she gets factory service and no doubt a right price. Her Lil Otto turbine runs on a working pressure of 18 pounds per square inch generated by about 40 feet of head. She uses a 9/32 inch nozzle, consuming about 9.7 gallons per minute, to produce an output of 1.35 amperes.

Joyce's turbine is producing about 18 watts, with a daily output of 430 watt-hours. During the winter, Joyce's PV panels produce a daily average of about 336 watt-hours because the mountains shade them most of the time. So during the winter, Joyce's nanohydro turbine produces more electricity than the PV array (even though the PV array has a peak output almost ten times greater). The third power source is an aged two-cylinder diesel engine/generator. When all else fails, Joyce can fall back on the generator to source her system.

This six kilowatt generator also sources the machine tools in the Lil Otto Hydroworks building nearby (more on this later). Joyce uses lead-acid batteries, housed in her basement, to store the power her PVs and nanohydro produce. She uses two Trojan L-16s, a 200 ampere-hour, 12 volt Interstate diesel starting battery and an assortment of other 12 volt batteries. Total capacity of her battery pack is about 500 ampere-hours at 12 VDC. Joyce's system also uses a brand spanking new Trace 2012 inverter to convert the DC stored power in the batteries into 120 VAC for her appliances. Also located in the basement is a Heliotrope CC60 PV charge controller that rides herd on the array's output. I asked Otto why he had such a large (60 ampere) control on the 12 amp array. Electrically speaking, Joyce has all the conveniences. For example, her refrigerator/freezer is a super-efficient, 12 VDC operated, SunFrost RF-12. This refrigerator/freezer consumes about 290 watt-hours daily in Joyce's kitchen. Otto is busy taking data on the SunFrost's performance with a motor run-time meter. Joyce's home is primarily wired for 120 VAC, but there are a few special 12 VDC circuits directly supplied by the battery. Joyce uses 12 VDC for a fluorescent light on the ceiling of the kitchen, for her CB radio and for the SunFrost. Entertainment electronics are powered by 120 VAC from the Trace inverter. Joyce runs her washing machine when Otto's out in the shop and the large generator is operating.

The Lil Otto Hydroworks

We also visited the shop that Bob-O and Otto use to make their nanohydro turbines. Against a background of machine tools, ranks of Lil Otto turbines march down their assembly line, jump into boxes and travel to streams and springs round the world. It was inspiring to see the obvious care and thought that goes into their manufacture. Bob-O and Otto start out with a permanent magnet Bosch generator. This generator is coupled to a molded turbine wheel made by "Powerhouse Paul" Cunningham at Energy Systems and Design in Canada. The generator is housed in a sealed PVC pipe case. Bob-O and Otto are now installing a new *gravity tube* along the shaft of the unit to eliminate water infiltration to the generator's innards. The unit is supplied with a blocking diode (to keep the generator from becoming a motor) and filtration to keep electrical noise from interfering with radios and TVs. There is a 0-8 ampere output meter on Lil Otto's top so operation can be checked at a glance. The Lil Otto units will produce up to five amperes, with enough head and flow. Where this turbine really shines is in the gallons per minute required for operation. This turbine consumes very little water. For performance data on these turbines, see the *Home Power* #13 "Things that Work!" article about Lil Otto. Bob-O and Otto deserve credit for intelligent and efficient use of off-the-shelf components in manufacturing

Lil Otto turbines. For example, the housings are sections of stock PVC pipe. The various sized nozzles used (and there is one to fit every site) are stock Rainbird™ sprinkler nozzles.

Well, we were ready for more. Obviously Otto's mom, Joyce, was satisfied with her PV/Hydro system and was justly proud of her inventive son. But she's Otto's mom and could be biased. We asked to talk to some paying customers. Bob-O smiled and invited us on a trip down river.

Getting Down River

This turned out to be an adventure in itself. The road that winds along the Salmon from Cecilville to Forks of Salmon is mostly one lane with sharp 100+ foot drops into the surging river. Kathleen drove first because she had a CB radio in her car. The CB radio is essential because you have to know when a logging truck is coming so that you can pull out in a place that is wide enough for the truck to pass. Kathleen (also a ham radio operator) kept us advised of traffic on our two meter ham radios. As I drove along I had trouble keeping my eyes on the road; the scenery was too distracting. Rock cliffs plunged down into the foaming river. From bends in the road, large mountain meadows filled with trees soothed my senses. I like mountains and the peace they give. The Salmon Mountains are very beautiful. It is easy to understand why these folks live in such a remote place.

Terry and Betty Ann Hanauer's Home

After about 30 minutes of driving we arrived at another PV/Hydro system at the home of Terry and Betty Ann. Betty Ann, a school teacher, took time off to show us her well-built and immaculate home. This large, owner-built home houses their family of six people. Their home has been powered by site-generated electricity since 1987.

Power sources at the Hanauer home are much the same as at Joyce Eichenhofer's home. The Hanauers use a PV array composed of three 36 watt Solavolt panels. On an average day this array makes about 600 watt-hours of electricity. These panels are fortunately located on one of the sunnier locations along the river. Terry and Betty Ann also use a Lil Otto turbine. This Lil Otto, however, is located at a much better site than Joyce's. At Terry and Betty Ann's site the turbine has 72 feet of head to work with (32 pounds per square inch dynamic pressure). Here the turbine produces 2.5 amperes with a ¼ inch diameter nozzle consuming ten gallons per minute. Terry and Betty Ann's turbine produces 33 watts and makes 810 watt-hours of electricity per day. Combined production of both the PV array and the nanohydro turbine is about 1,400 watt-hours daily, and that's enough to run a household with four kids! Terry and Betty Ann also use an engine/generator (Onan two cylinder six kilowatt powered by propane) for extended cloudy periods and times of intense power consumption. Terry and Betty Ann use a battery pack of four Trojan L-16 batteries to store the power produced by Lil Otto and the PV array.

This battery pack is housed in an insulated blister on the outside of the house. Bob-O Schultze fabricated a custom regulator for the PV array. A Trace 2012 inverter with 110 ampere battery charger is used to power the house and recharge the batteries when the generator is running. Betty Ann says that with four kids, the washing machine gets a lot of action. She starts the generator, does the washing and refills her batteries all at the same time. The Hanauer's home is wired for 12 VDC lighting, which spends a lot of time operating. The Sabir refrigerator/freezer is powered by propane. Betty Ann is a gourmet cook, and her kitchen is filled with good things. Among these things are many kitchen appliances (food processors, grinder, mixers, blenders and such) that all run from the inverter. Betty Ann said that cooking was more enjoyable because she didn't have to start the generator just for a few minutes of kitchen appliance use. Everyone likes not having the generator yammering while reading or listen to the stereo. Hot water is produced by a large solar collector located next to the PV array. Betty Ann told us that in the summer, even with wash and four kids to bathe, there is more than enough hot water being produced by their solar collector to meet their needs.

Lessons Learned

From experience, the folks along the river have learned a great deal about making their own power. They've learned that even a trickle can be turned into a watt. They learned to use a variety of natural power sources without damaging their environment. And certainly, they've learned contentment and happiness.

Editor's note: This article first appeared in *Home Power* Issue #15 (February/March 1990) and is reprinted with permission of the author.

Mini Hybrid Power System

R. T. GAYDOS AND LINDA GAYDOS

Eight years ago we purchased remote acreage in the Sierra Foothills of Northern California and wished to build a small energy efficient cabin on it. After discussion with the local power company we were told that it would cost US$10,000 to extend electrical service. Our cabin site is 1,800 feet from the closest power pole, and they charge about US$5.60 a foot. Also, the power company could not even give us a possible date as to when we could expect electrical hookup. The only affordable solution was to make our own electricity by alternative means. We were fortunate to live in an area where there were two renewable energy stores, so ideas and advice were readily accessible.

Hydroelectric

In our area the annual rainfall is 55 inches, and runoff from Sierra Mountain snow melt creates an abundance of natural spring water. Our springs are well above our cabin site, so we've got gravity flow water. This coupled with the fact that a small inexpensive 12 volt hydroelectric generator was being made locally made the decision to go hydro easy. ·

The main component in our system is the HydroCharger I™, designed and built by Sam Vanderhoof of Independent Power Company in North San Juan, California. This is the smallest hydro unit I know of. It weighs approximately 20 pounds and is 12x8 inches in size. The hydro unit has a four-inch diameter Pelton wheel rotating on the horizontal axis, connected to a small permanent magnet generator. According to the manufacturer, it will generate up to two amperes of current and begins producing electricity with water flows of 12 gallons per minute at three feet of fall.

We have about 40 foot fall with 500 feet of run, going from water source (springs) to the cabin where the generator is located. The two springs we have tapped give us a total of six to 12 gallons per minute of flow depending on the time of year. This gives us from 1 to 1.5 amperes from hydro. The hydro runs constantly, therefore we get 24–36 ampere-hours per day. In our system water is collected in a catch basin under each spring, from where it runs into a 35 gallon plastic reservoir and through 1.25-inch PVC pipe downhill 500 feet to the cabin. Larger pipe, at least 1.5 inches, should have been used to reduce interior line resistance. The hydro unit is located under the floor of the cabin with a valve running up through the floor of the bathroom, allowing the hydro to be turned on and off inside. This is helpful because the gravity flow water system that feeds the hydro unit is also our domestic water supply. When we are getting low flows of water, like late summer and fall, we can turn off the hydro unit to obtain more water pressure for showers.

Occasional cleaning of the debris from the catch basin's screens is the only maintenance needed for the water collection system. The only maintenance adjustment we need to make on the hydro unit is to change the water nozzle size depending on available water, i.e. too big a nozzle without enough water will cause air to be sucked into the water line (penstock). Nozzle orifice sizes we use are ¼ inch to ½ inch in diameter.

We originally had trouble with the Hydro-Charger I™ and replaced the lower bearings in the generator unit several times. Acidic water was getting inside the generator and causing severe corrosion. This problem was solved by the manufacturer, by elevating the generator from the housing with ¾ inch standoffs and putting a splash guard collar on the generator shaft. This modification is now standard. Since last bearing replacement and modification, the HydroCharger I™ has run continuously for four years without any trouble. It is a very reliable component.

Photovoltaic

Another part of our mini hybrid power system is a single Solex 35 watt PV panel; it produces six to ten ampere-hours per day. The panel is located on the roof of the cabin and is accessible by a roof ladder for cleaning, snow removal and redirecting toward the sun. Although roof mounting is not best, it was the only place close to the cabin that would give optimum solar exposure. The PV panel was also purchased to qualify the system for the now defunct State and Federal Solar Energy Tax Credit.

Backup Power

The third producer of this hybrid system is a used 2,800 watt Yamaha gasoline engine/generator. It is used in conjunction with a Sears 50 ampere heavy duty battery charger to charge the 12 volt battery bank when it gets too low. It is needed infrequently to charge batteries, but is necessary to run a ten-inch radial arm saw and Maytag clothes washer.

Energy Storage — Battery Bank

The 12 volt power produced by hydro, solar panel or generator/charger is transferred and stored in a bank of deep cycle, lead-acid golf cart batteries. There are six, six-volt, 220 ampere-hour batteries wired in parallel and series to yield a 660 ampere-hour, 12 volt storage bank. Batteries are located inside the cabin in a window seat/battery box. The box is sealed to the interior of the cabin and vented through the wall to the outside. The vent dissipates the hydrogen gas created by the batteries.

We have used the same collection of batteries for four years. Unfortunately, all six batteries were not purchased at the same time, i.e., the first set was used, the second set was purchased nine months before the third set. This was definitely an error because batteries develop a charging memory and will only accept a charge as high as the oldest or worst battery. None the less, they are forgiving because the batteries are continuously being charged by the hydro unit. If we were depending primarily on solar panels, we doubt that the batteries would be so forgiving. See *Home Power* Issue #9, page 27, for more information on lead-acid batteries.

Distribution and Consumption of Power

From the battery bank, 12 volt power goes through a DC circuit breaker panel to its various points of consumption. We have eight separate 12 volt circuits. Twelve volt battery power also supplies the Trace 1512 inverter. The inverter's AC power then goes through an AC circuit breaker panel and on to various points of AC consumption. There are four separate AC circuits, with one going up to the woodworking shop which is 100 feet away from the inverter.

Appliances run by 120 VAC inverted power are a small microwave, phonograph turntable, word processor, toaster, blender, mini drip coffee maker, vacuum cleaner, hair dryer, hair rollers, small clothes iron, eight-inch table saw, 7½ inch Skil saw, drills, sanders and ten-inch miter saw. The only thing the inverter will not run is a Sears ten-inch radial arm saw and an automatic clothes washer. These must be run by the AC gas generator. All AC appliances were purchased with energy efficiency and low power consumption in mind.

The main consumer of power in our system is 12 volt lighting. We have 14 separate lights, which are either incandescent, fluorescent or quartz. Fluorescents seem to be the most efficient (illumination per ampere) with quartz a very close second and incandescents a distant third. We have strategically placed lights and semi-gloss white walls to help reflect light. Location is the most important factor in efficient lighting. Lighting uses approximately half of the power we make. Our other 12 volt appliances, 12 inch black and white TV, hi fi, CD player and fans use another quarter of the power generated. The remaining quarter of the 12 volt power is inverted to 120 volts AC and is consumed by AC appliances.

We have on the average 36 ampere-hours at 12 VDC per day of power available, depending

System Costs

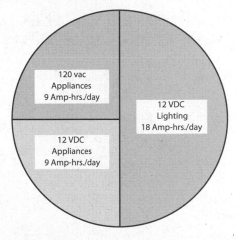

Home Power

System Costs	
HydroCharger I™	$500.00
35 watt Solex panel	$300.00
2,800 watt Yamaha gas generator (used)	$400.00
Six six-volt DC batteries (2 used)	$300.00
Trace 1512 inverter	$1,100.00
500 feet of 1 ¼-inch PVC pipe	$200.00
Monitoring panel, circuit breakers, wiring, outlets, how to books and publications, fans, light fixtures and miscellaneous	$1,500.00
Total equipment cost	**$4,300.00**

Note: Costs in US dollars

on hydro and solar panel output. Power is monitored via a metering panel which tells voltage of battery bank and amperage being consumed. Amps and volts output of hydro and solar panel are also displayed. The metering panel was built with analog meters and is flush mounted in the wall above the circuit breaker cabinet. It is helpful to see what's happening via gauges, especially with the hydro charger, because its current output can easily be translated into water output (gallons per minute).

There is no controller because amps produced are small enough that the battery bank cannot be overcharged.

The actual cost was greatly reduced due to the State and Federal Solar Energy Tax Credit which saved us approximately 40%. This incentive to save energy and use renewable energy sources has sadly expired. The cabin, which is 600 square feet, was designed and built with this power system as an integral part. It took us about 400 hours to design and install all electrical components of the system. I would venture that a professional could have done it in half the time.

Ideas and Ramblings

Our energy needs are also met by using propane for refrigeration, cooking and hot water. We have an Aqua Vac on-demand hot water heater, supplemented by a small water heater in the woodstove. We use approximately 200 gallons of propane a year. Our Thelin Thompson T-1000 woodstove is thermostatically controlled by a 12 volt freon damper switch and is the winter space heating source for the cabin.

The US average power usage is approximately 10,000 watt-hours per day. By being conservative and designing a small home's lighting and electrical needs efficiently, we manage to be comfortable on approximately 700 watt-hours a day. In the future we plan to install another HydroCharger I™ downhill from our cabin and recycle expelled water from the first hydro unit to operate a second unit. It will have 40 feet of fall and be 140 feet away. It is estimated that power generation will increase 50% to 60%; this will enable us to run an efficient automatic clothes washer and color TV.[1]

Editor's note: This article first appeared in *Home Power* Issue #11 (June/July 1989) and is reprinted with permission of the authors.

Mini Hybrid Power System

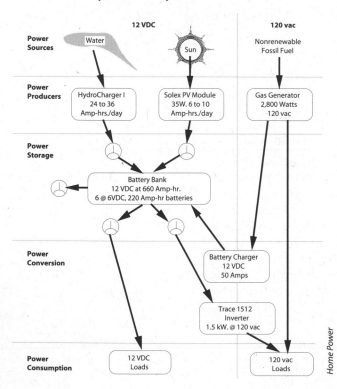

Home Power

18

Hydro Power Done Dirt Cheap

STEPHEN M. GIMA AND EILEEN PUTTRE (LOSCHKY)

Eileen and I are both firm believers in the information superhighway. She uses it (via the Internet) and I build it, being employed by a telecommunications company. Maybe it was ironic when we started looking for a home in the Adirondacks, we fell in love with the one a mile off the grid.

It's a log cabin, built by a local mason, on an abandoned logging road. I guess after the initial construction in 1980, he kinda lost interest. When we found it, it looked like it was hardly ever used. The center hall fireplace and stove could not have burned more than a ¼ cord of wood. While the house had a kitchen and bathroom, the water fixtures only got water in the spring, due to the rise in the creek behind the house. There were also no lights except for the camping lantern we used to bring up. About the only thing that did work with any regularity was the propane oven.

Since light was our first consideration, we discovered by thumbing through non-electric catalogues that Humphrey made wall-mounted gas lights. So with 100 feet or so of ⅜ inch copper tubing, the main living quarters, downstairs, now had lights.

Well, about this time Eileen got a corporate level job with a photovoltaic company.

We were thrilled! Maybe we could actually produce our own electricity. Our euphoria didn't last. Even at cost, photovoltaic panels were pretty expensive and for that part of the country especially, since it seems the sun hardly ever shines. Upon further investigation, it seems we get the least amount of available sunshine in the lower 48 states.

By now we were learning a little about renewable energy. The creek turned out to be a gold mine. Searching through and thoroughly reading everything we could find on

Steven Gima working on the final assembly of his $328 hydro system.

Eileen Puttre (Loschky)

the subject, we became convinced that a microhydroelectric system was the way to go. But still, a Harris Hydroelectric generator (at about $1,000) was still a little more than we could afford.

A friend of ours in the Adirondack area, who happens to be an electrician, thought we might try to build a hydroelectric generator ourselves. Our friend located an American Bosch 12 VDC permanent magnet generator through a surplus catalogue.

The phone company, where I work, was trying to unload 6,000 feet of reeled 1½ inch semi rigid conduit used for buried fiber optic applications. Over the next few months I managed to get about 1,000 feet of it in roughly 200 foot coils. Try coiling 200 feet of 1½ inch semi-rigid conduit, then hauling it in a Toyota pickup 200 miles. It's a wonder that we never got stopped by the police; maybe they just shook their heads and laughed. But we never had any trouble. As hard as it was coiling the conduit, uncoiling it was even worse.

We ran the first piece from a dam we rebuilt (twice) in back of the house, along the creek bed and down, an overall drop of about 35 feet to what looked like a suitable spot to secure our little hydro setup. With stop watch and buckets in hand, we determined the flow through the conduit to be about 35 gallons per minute. Over the course of the next several weeks, we ran a total of three 1½ inch permanently lubricated semi-rigid conduits, each 265 feet long. We placed ball valves half the distance from the dam to the generator.

The dam's been rebuilt (the mason who built the house constructed the original dam) placing a six-foot length of six-inch PVC on the bottom, then grading on an incline with rocks. It has only washed out once since then, but we've learned a lot about dam building. The three conduits were drilled and screened with 30 opposing ½ inch holes along the length at the dam then pushed through the six-inch PVC and secured.

The creek flows from behind the house to around the side about 100 feet from the house. The hydro system, about 150 feet from the front of the house, was set up on a rock stand next to the creek and secured in place with cement. The three 1½ inch conduit pipes were glued to two-inch sweeps aimed at a Pelton wheel and reduced to ⁵⁄₁₆ inch nozzles. All this took an entire summer of weekends. By mid-October we were ready to test. Without even owning a multimeter at the time, we took an old automotive headlight wired directly to the generator, turned on the valves and surprise, surprise, it lit. I doubt Thomas Edison was as happy as Eileen and I. We happily danced and congratulated ourselves for hours. By that evening we had run two #6 AWG wires up to the rear of the house and hooked up our headlight direct. The entire back area of the house lit up. From then until mid-December, when snow made it impossible to get to the house by car, we would go up for weekends, open the valves and turn on our light.

So far we had spent about $80. The generator was only $13, and the ball valves were about $20 each. I made the nozzles from a box of spare plumbing parts. The six-inch conduit I found. The 800 feet of conduit, the PVC sweeps and the squared and hollowed tub for the hydro plus all the wire (considered scrap) was courtesy of "Ma Bell."

Our electrician friend had mountains of old electrical switches, fuses and boxes. We told him what we thought we might need, which he gave to us. We went home for the winter and started to clean and separate everything.

When we started all of this, I knew virtually nothing about DC electricity, but by spring we had put together our pull-out fused disconnect with two 60 amp cartridge fuses and our fused DC load center pieced together from several old glass buss fuses and holders. My son, Jesse, had a five year old battery in his car, so he got a new battery and we got his old one. Don't laugh, it worked. So that spring we were ready to make our system as safe as we knew how and to bring electric lighting indoors. We cut out a spot under the living room steps for access to the crawl space below the house. Luckily we chose that particular spot. We had about two feet from the floor joists to the dirt below. The rest of the crawl space wasn't so spacious, but being somewhat thin, I managed to fit.

Having access to an unlimited supply of #6 gauge wire, we bonded two pieces twice for positive and negative. This is roughly the equivalent of #3 gauge wire. It is well within line loss limits for the 150 feet from the hydro system to the house. The charge controller, main disconnect and DC outlet center are all set up under the steps. The batteries are directly under the steps in the vented crawl space next to the access door.

The automotive battery plus two 12 VDC Hawker Energy HD30 batteries worked well all summer. For the winter, the batteries remain home in New Jersey along with our dump truck and bulldozer batteries. All are kept in the garage on solar chargers. We'll probably get two six-volt DC golf cart batteries for next

spring, but we were pleasantly surprised that the old car battery performed so well.

So far we've only been able to get the hydro to put out 1.75 Amps, but it's enough to keep the batteries charged. We shut down the system during the week while we're not there and turn it on Friday night until Sunday morning.

What started as a headlight burning in the back has grown to be lighting for a tool shed,

Top: A close-up view of the Pelton wheel (five inches in diameter) and the two two-inch sweeps that end in improvised ⁵⁄₁₆-inch nozzles

Bottom: The completed hydro plant showing the two diverters used when the cabin is unoccupied.

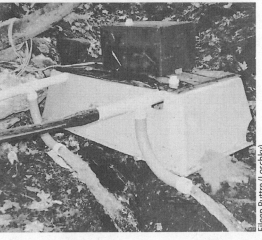

Eileen Puttre (Loschky)

woodshed, front porch, bathroom and, soon, upstairs bedroom. The downstairs is still using gas lights and they're great, each producing the equivalent of 50 watts of light.

But the biggest benefit is the electric water pump. We pump water from the creek to the tool shed where the pump and propane water heater are located. From there, it is another 60 feet to the house. Finally, last summer and fall we had indoor hot water showers. Until then, we used a solar shower on the front porch, which is fine in July but a little tough around

Steven installs the gravity feed water system which is filled by the hydroelectric powered pump.

Stephen and Eileen's Hydro System Cost	
Plastic Pelton wheel	$75 (23%)
5.5 amp, 12 VDC, PM Generator	$13 (4%)
SCI Mk III Regulator with meters	$110 (34%)
Three ball valves	$60 (18%)
Battery safety switch	$20 (6%)
Lugs, fuses and hardware	$50 (15%)
Total Cost	**$328**

Note: Everything else for the hydro system was either found or donated.

October and November. Good thing we're a mile from our nearest full-time neighbor. We've only had a few close calls while showering on the porch. Luckily you can hear approaching visitors before they see the house.

Since we only use the house on weekends and a few weeks during the summer months, all the appliances are 12 volt DC models. We thought of adding a small inverter but we get along just fine for now.

Acquiring the knowledge and resources for our place in the woods was an enjoyable learning experience and one we hope to duplicate when it comes time to build our permanent home in the Adirondacks.[1]

Editor's note: This article first appeared in *Home Power* Issue #52 (April/May 1996) and is reprinted with permission of the authors.

System Layout

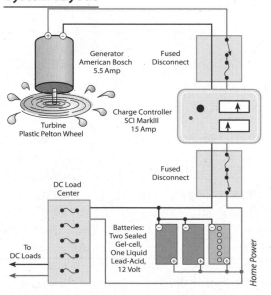

A Microhydro Learning Experience

LOUIS WOOFENDEN, WITH JO HAMILTON
AND ROSE WOOFENDEN

The potential for microhydroelectric power can't be beat the way it rains in the North Cascade mountains of Washington State. The climate in the foothills is wet, and many trees are covered with thick moss. It is cloudy much of the year, and though solar and wind are options, hydro is more cost-effective for those with a good stream. At Floyd and Frasia Omstead's property in the Skagit River valley, hydro is the best renewable solution. Although grid power is close, the Omsteads prefer to have the independence and reliability of a hydroelectric system.

Ian Woofenden (my dad), Solar Energy International's (SEI) Northwest coordinator, approached Floyd with the idea of installing a hydroelectric system as part of an SEI workshop. They had met two years earlier when Floyd attended an SEI PV workshop. After considering the pros and cons, Floyd agreed to provide the installation site for the SEI microhydro workshop.

The workshop was held in October of 1999. A class of students spent five days learning about microhydro. They came from the Northwest and beyond, including Montana, Pennsylvania and California. There were students planning to work in the renewable energy industry, others who wanted to build their own hydro systems and one student who had concerns about the Y2K computer bug. There were also those who wanted to learn more about the systems they were already living with.

Hydro Education and Experience

The first two days were dedicated to classroom instruction taught by Bob Mathews from Energy Alternatives of British Columbia, Canada. Bob is very knowledgeable about microhydro and has designed many systems in Canada and in the US. His organized and detailed approach gave the students a real grounding in microhydro theory and design principles.

On the third day of the workshop, we toured a couple of hydro systems, led by SEI instructor Johnny Weiss. First we stopped to see Chris Soler, who has a small homebrew system with a Harris Pelton wheel coupled to a salvaged car alternator. His system showed what you can do with a very limited budget and a tiny stream. We also toured the Canyon Industries manufacturing plant in Deming,

Washington. Dan New, owner of Canyon Industries, showed us around their production shop. We saw turbines of various sizes that were going to be shipped all over the world — one was going to Honduras, another to Montana.

Dan has a 25 kilowatt AC system that provides some of the power for the facility. We toured the system from the intake to tailrace and talked with him about the special problems of AC systems. It was fun to see a larger system in action.

Down to Business

Thursday morning, the SEI class carpooled to the Omstead property. After everybody arrived introductions were made, and the class toured the site. We found that Floyd was not as ready as he had agreed to be. The ditch was not completely finished. The power shed was full of stuff. The platform for the turbine wasn't painted. We also later found out that there wasn't enough pipe or enough fittings.

We had to work hard just to finish on time. This was good in some ways, because it showed the students how to do an installation quickly. After finding out how much there was to do, we broke into three groups and began installing the system. The intake, penstock and electrical system were all tackled by different work crews. We did finish, with the group's hard work, and a truck full of tools and parts brought by SEI alumnus John Klemmeson who came to help out.

We Capture the Water

The intake was designed as an in-stream diversion — a small dam in effect. This means that the intake blocks the whole stream and creates a small pond. The intake structure was made of pressure-treated 2×12 lumber, with black plastic sheeting sealing most of the cracks. The diversion was divided into two sections, the intake section (which was screened) and the spillway section (where the extra water spills over and returns to the stream).

We used the small dam because local residents said that not much silt or debris comes down the stream in the winter. Just above this system is a system with three overshot waterwheels, and we expected that it might filter out whatever did come down. Now that the system has been in operation for a few months, it's obvious that we were all mistaken. Floyd reports that the impoundment area fills with silt and rocks, and he has to clean the screen quite frequently. He is now in the middle of designing a better intake system to deal with the actual stream conditions.

The penstock was attached by first cutting four-inch holes in the pressure-treated lumber. Then a short piece of pipe was glued to two couplings, one on each side of the wood. We then had a *coupling sandwich*, and it was secure. Next we glued on a tee, to which we later attached an air vent.

The air vent prevents the penstock from damage if the intake is suddenly clogged. If there were no air inlet and intake was clogged, the force of the water rushing down the penstock could cause a vacuum that might collapse the penstock. We also added two cleanouts low in the dam wall.

Penstock

When we arrived at the site, the 450 foot long ditch to bury the penstock was mostly completed. It was dug by a backhoe, and it needed to be evened out and dug a bit deeper in a few places. While some of us worked on the intake and other projects, a few of the students and staff did the required ditch digging. After the ditch was finished, it was about 18 inches deep.

The crew then laid sticks across the ditch about every five feet to hold up the penstock, so that it was easier to glue the pipe together. We removed the sticks and lowered the pipe into the ditch before we pressure tested the penstock. The penstock was constructed of 10 and 20 foot sections of four-inch ABS pipe.

Manifold and Turbine

The manifold was built from four-inch ABS pipe. At the bottom of the penstock, there is a tee. We connected one side of the tee to a two-inch ball valve. Ball valves were not the best choice, but gate valves in this size are expensive. A ball valve can be closed very quickly. When you close any valve too quickly in a hydro system, you create a surge of pressure at the bottom of the penstock. This is called water hammer. Water hammer occurs because the water rushing down the pipe has a lot of force. If you suddenly cut off an outlet for the water to escape, then the water that is moving pushes the water in front of it. Because water isn't very compressible, a lot of pressure builds up. This can burst joints of pipes, or the pipes themselves. If Floyd had bought gate valves instead of ball valves, the potential for water hammer would have been reduced, because a gate valve takes more time to close than a ball valve.

After the ball valve came a nozzle. We cut the four nozzles to different diameters, so that Floyd has more options for how much water to use. The other side of the first tee went to a 90° fitting. After the fitting came a short section of pipe, which was connected to another tee. One side of this tee went to a valve and nozzle, and the other side continued to another 90° fitting. The manifold continued in the same way to the other nozzles. It was quite a puzzle to get it together properly, and we had to do some cutting and refitting.

Stream Engine

Floyd chose the Energy Systems and Design (ES&D) Stream Engine, which is made in New Brunswick, Canada. The ES&D turbine is a bronze Turgo wheel coupled to a permanent magnet alternator. The permanent magnet alternator requires less maintenance than an alternator with brushes, but it is harder to adjust for maximum output.

Turgos can handle more water than Pelton wheels. At Floyd's site, this is a definite advantage, because of his high flow in the rainy season. The head (vertical distance from the intake to the turbine) is only about 30 feet, and the lowest flow is around 40 gallons per minute. But in the winter, this flow can more than quadruple. Floyd wanted to be able to handle as much water as possible, so he chose the Turgo runner.

Electrical System

The electrical system was installed in one side of a small shed about 20 feet from Floyd and Frasia's cabin. The shed was partitioned, with one side for the electrical equipment and the other side to be used as a wash house. Floyd chose to have only AC loads, so he will be running all his loads through his Trace inverter. Because he is not running any DC loads, his choice of a 48 volt system makes good sense. A higher voltage allowed us to use smaller wire in between the turbine and the power shed, because of the lower line losses.

John Heil from Dyno Battery headed up the electrical crew. He provided expert instruction and design assistance even while under the stress of working with his competitor's batteries. By the end of the first day, the inverter and batteries were wired, and the inverter was providing AC power from the energy stored in the batteries. The batteries are eight Trojan L-16s and are connected in series to provide 350 amp-hours at 48 volts DC.

We used ¾ inch plywood for a backboard to mount the gear on the wall of the power shed. The crew mounted the Trace 5548 sine wave inverter, DC disconnect box and Trace C-40 charge controller on the wall. The AC output from the Trace was connected to a multi-plug outlet mounted on the wall of the powerhouse, since Floyd wasn't ready to run it to the cabin.

Hydro

We also installed a Cruising Equipment E-Meter to monitor the system. The C-40 prevents the batteries from being damaged by overcharging. We ran the output of the turbine through a 20 amp disconnect. This prevents the system from damage if a short or other fault occurs. We then wired the C-40 to dump any excess power into a heating element attached to the wall. The heating element will provide some heat to the battery room and wash house.

Earlier, some of the class had dug a trench between the power shed and the turbine site. The electrical crew laid PVC conduit in the trench. Then they pulled the wires through the conduit, and we were almost ready to test the system.

Putting It All Together

Most of the separate parts of the system were now complete. The electrical system was ready to accept the power from the turbine. The intake was all prepared, and the manifold was assembled. All we had to do to send water down the penstock was to put in the cleanout plugs in the dam. But down at the turbine site, there was still work to do. We needed to connect the manifold to the turbine, install a pressure gauge, test for and fix any leaks and wire the output of the turbine.

It was quite a job to attach the manifold to the turbine, and not without its problems. To get the manifold on, we had to flex the pipe slightly. After attaching most of the nozzles to the manifold, we tried to bend it too much in an attempt to get it all together, and we broke a nozzle.

Luckily, ES&D provides six nozzles with their turbines, so we just replaced the broken nozzle and began again. This time we put it all together with no mishaps.

We added a wye to the end of the penstock above the manifold. From the bottom of the wye we continued the pipe in a ditch, all the way to where the water from the turbine runs back into the stream. On the end we attached a cleanout. Floyd can periodically open the cleanout and let any residue from the bottom of the pipe wash back into the stream. We connected a short piece of pipe to the top of the wye and connected a tee to it. We glued an adapter and valve to one side of the tee. This is so Floyd and Frasia can tap water from the penstock for irrigation and fire protection.

We glued a 22½° fitting on the other end of the tee, and a piece of pipe about three feet long. This brought us almost to the manifold. We added one more 22½° fitting and glued it all together. We then drilled a small hole in one of the fittings and threaded the pressure gauge in.

Ready to Rumble!

Then…the time we'd been waiting for. We were ready to pressure test the system. Someone ran up to the intake of the penstock and closed the cleanouts. We left the irrigation water valve open for a while to let the air escape. In a couple of minutes, the penstock was full of water, and the pressure gauge read 16.5 pounds per square inch. We wired the output of the turbine to the wires running through conduit to the power shed.

Now we were ready to test run the turbine. One of the students held a multimeter, which was set up to measure how much electricity was being produced. The Stream Engine doesn't have an ammeter included. Instead, inside the connection box is a shunt and two test points. It's set up to show you the output amperage by measuring DC millivolts at the test points.

Someone opened the valve. The turbine spun up to speed. It was making power! The meter read 5.8 amps, which is about 280 watts at 48 volts. At this point we all had big smiles on our faces, and we were gathered around, looking at the meter and listening to the hum of the turbine.

We weren't done yet though. We had started the turbine using the largest nozzle, but that soon brought the water level at the diversion down. If we continued to use the water

at that rate, pretty soon we'd end up with air in the pipe. If any air is in the penstock, the turbine will still run, but you lose head and therefore power.

Adjusting Flow

One of the students had a pair of FRS transceivers. So while the turbine crew turned on a nozzle, I watched the water level at the diversion and reported back to them on the radio. If the water level went down, that meant that we were using too much water, so we tried the next smaller nozzle. We installed the system in the season with the minimum flow it will see. In the winter, spring and early summer, there will be much more water, and Floyd will be able to use more nozzles and have more power output. We found that even using the smallest nozzle, the water level was still going down, so we replaced one of the nozzles with a new nozzle cut to a smaller diameter. That did the trick. The water level didn't fall. It actually spilled over the top of the dam a bit, which is just what we wanted.

Adjusting Output

We started working on adjusting the Stream Engine for maximum output power with the smallest nozzle. The Stream Engine has magnets that spin, and fixed coils. The magnets can be moved up and down, nearer and farther from the coils. For any revolutions per minute, the closer the magnets are to the coils, the more power is produced. But if the magnets are moved closer, the turbine slows down. To produce power, the turbine has to spin above a certain number of revolutions per minute. The trick is to find a balance between having it too slow and not producing any power, and having it too fast (magnets too far away from the coils) to produce too much power.

We didn't have time to do the process thoroughly, but we got the magnets adjusted to approximately the right distance. Just doing it roughly took a long time. First we had to close the valve, and wait for the turbine to stop. Then we had to loosen a nut that locked the magnets at whatever distance they were from the coils. Then we screwed the magnets closer to the coils. Then we had to tighten the nut that locked it. Finally, we opened the valve and let the turbine spin up. Then we checked what the power output was. If the power had increased from the last test, we repeated the process all over again and moved the magnets even closer to the coils. If the power had de-creased, we backtracked and moved the magnets farther away. After everything was adjusted, the turbine put out 2.1 amps at 48 volts, or about 100 watts.

Useful Energy

By this time, it was dusk. We finished up the last minute details and headed up to the power shed. We plugged in a light and looked around at our handiwork. We looked at the E-Meter, and saw what the voltage and amperage were. The class went outside, and gathered around. Goodbyes were said, and addresses exchanged. The students drove off into the darkness, tired but satisfied to know that the job was done, and that they had a good working knowledge of how a microhydroelectric system works.

This system will make a big difference for Floyd, Frasia and their two children. Floyd reports that for the past two months, he's been

After two long days of work, remaining SEI students, staff, family and friends gather for a group photo with running turbine.

Ian Woofenden

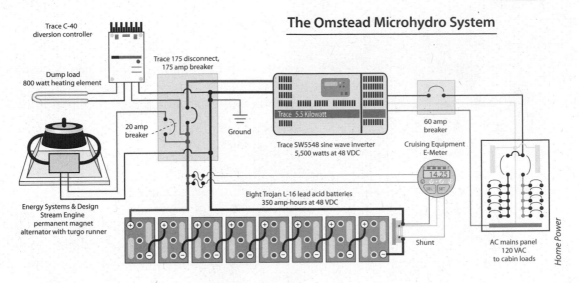

The Omstead Microhydro System

Trace C-40 diversion controller

Dump load 800 watt heating element

Trace 175 disconnect, 175 amp breaker

20 amp breaker

Ground

Trace 5.5 Kilowatt

Trace SW5548 sine wave inverter 5,500 watts at 48 VDC

60 amp breaker

Cruising Equipment E-Meter

14.25

Energy Systems & Design Stream Engine permanent magnet alternator with turgo runner

Eight Trojan L-16 lead acid batteries 350 amp-hours at 48 VDC

Shunt

AC mains panel 120 VAC to cabin loads

Home Power

running the turbine on three nozzles with an output of 11 amps at 48 volts, or over 525 watts continuous. This gives the Omsteads a significant energy surplus, so they don't have to worry much about what loads they use during the winter. The hydro system has allowed them to move into their cabin and stop paying rent. It will provide them with ample power year-round. When wind storms take out grid power, it will keep their lights shining when others are lighting candles. If efficient loads are used, it will provide power for them to eventually build and electrify a larger house. The system will help people, which after all, is what renewable energy is all about.[1]

Editor's note: This article first appeared in *Home Power* Issue #76 (April/May 2000) and is reprinted with permission of the authors.

Omstead System Loads

22 incandescent lights	33%
Fridge/freezer, 11 cubic feet	16%
Ceiling fan	16%
Chest freezer, 9 cubic feet	13%
Computer	8%
Toaster oven	5%
Two TVs and VCR	5%
Microwave oven	4%
Total average kilowatt-hours per day = 7.65	

Editor's Note: Incandescent lights use more than a third of this total. Better lighting would reduce the total loads to the 200 kilowatt hours per month needed for a high standard of living, and then some…

Louis Woofenden, with Jo Hamilton and Rose Woofendenv

Hydro—New England Style

BILL KELSEY

Lights make the spirit happy, especially here in New England, but I did not want to put up with the noise of a generator. I had been living off-grid using propane lights, which can be a fire hazard and are too hot in the summer. After talking with Gordon Ridgeway, a friend living off-grid on solar electricity, I began investigating alternatives. My property is heavily wooded, so I only get a few hours of sunshine on an average summer day, and even less in the winter. What I do have is a brook with a good downhill run. It drains about 2,000 acres, all in a state park with no houses.

My ten-acre woodlot in Sharon, Connecticut was once used by early settlers to grow fuel wood for their homes. Because of right-of-way restrictions, there is no access to the utility grid. With only a 30 foot drop and the cold New England climate to work with, I have constructed an innovative hydroelectric system that sustains me completely off-grid.

Starting from Scratch

When I started 20 years ago, I had never seen another hydroelectric system. I built mine completely by trial and error. I ordered a Harris Hydro turbine for 15 feet of vertical drop from a retailer. My first Rube Goldberg setup was a 275 gallon flat tank with the side cut out of it, so that water coming off the top of my rock and timber dam passed through a homemade filter and into two, two-inch pipelines laid side by side in the brook.

After many months building it and lots of friends' ideas, it was disappointing when the turbine only produced 30 watts at best. I tried to get a refund, but the supplier refused, so I was stuck with all my toys and no electricity. Out of desperation, I called Don Harris—and thank goodness I did! Don helped me save face. You see, I had become the laughingstock of the town. I love tinkering, so after Don explained the requirements of a successful hydroelectric system, I knew my brook could produce a substantial amount of electricity with the right setup.

Don started by telling me I had too much pipe friction, so I changed over to three-inch pipe. It worked a little better, but then I had problems with freezing in winter. Again, Don saved the day. He said I should measure the head (vertical drop) of the brook over the whole course on my property. That came to 30 feet, twice the head of my original system!

As luck would have it, my friend Bob was selling his 20 ton excavator for US$15,000, so I mortgaged my house to buy it. Next, I selectively cut oak trees on my property for cash to finance my project. We had a dry summer in 1999, so I hired a friend to run the excavator and we dug a pond and a trench running along the brook for 275 feet. I worked feverishly for three months burying four-inch PVC pipe 15 feet below the surface so it wouldn't freeze in winter. My welding skills came in handy when I fabricated the pipeline's intake, and I then built a concrete pad for the hydro turbine, complete with a ⅛ inch plate steel, bombproof cover.

In August, we sometimes have heavy rains. That August, we had 7 inches in 24 hours! I was scared to death, but my new earthen dam held, and I was off making 100 watts — not much, but you learn what you can live with.

One drawback in the early system was that Don's alternators loved to eat up brushes. I had a spare alternator around at all times so I could switch them out and replace the brushes, but it's a big job to change them. Thankfully Don came out with the brushless, permanent magnet alternator, which I've upgraded to. Now my hydro setup runs day in and day out, making a continual 150 watts. That's 3.6 kilowatt-hours per day, which I store in eight, six-volt golf cart batteries.

Steve Schulze of New England Solar Electric and Larry Riley, my electrician, helped me design my electric panel. This panel includes inverters that convert the DC energy stored in the batteries to AC electricity I can use to power standard, 120 VAC household appliances. It also has a remote start switch for a backup generator, which is located quite a ways from the house, since it creates a racket when it's running. A light at the house tells me if the generator is on. The Onan five kilowatt propane generator is used a little bit in the summertime when the stream's flow is low, but not much in winter — maybe 75 hours total per year. I use it primarily to power big electrical loads for my shop, which includes a welder and an air compressor.

The AC distribution panel feeds 120 VAC circuits, powered by the Trace DR2424 modified square wave inverter. I have a spare inverter just in case I have problems. A DC panel energizes DC compact fluorescent lights in each room of the house.

My system is all up to code. We have severe storms in New England, so even with the best engineering, the possibility of a direct lightning hit is still a threat. We have done a lot to protect from lightning, but there are no easy answers.

Using the Energy

With a frequent surplus of electricity, I decided to see how else I could use it. I had heard how great radiant floor heating is, but I wanted a DC circulating pump for the most efficient energy usage. As luck would have it, Ivan Labs Inc. was just coming out with a 24 VDC circulating pump — the El-Sid. Using my existing Aqua-Therm outdoor wood furnace for heating and the electricity from my hydro to run the pump, my home is peaceful and warm.

One thing for sure, Don Harris has changed my life forever — thanks, Don! Lights do make your life happier, and now I have more than enough electricity to run lights and basic appliances whenever I want. My wife and I lived in a 600-square-foot house for years. When

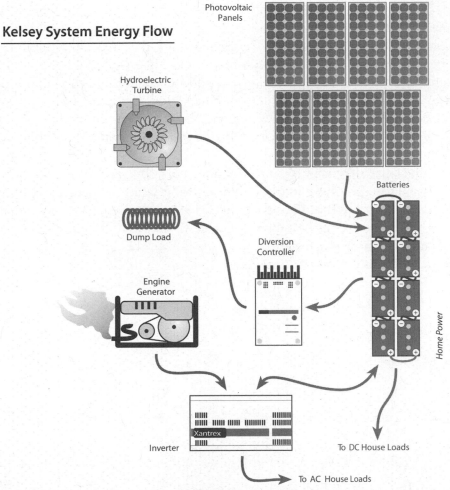

Kelsey System Energy Flow

Photovoltaic Panels

Hydroelectric Turbine

Dump Load

Engine Generator

Diversion Controller

Batteries

Home Power

Inverter

Xantrex

To DC House Loads

To AC House Loads

Note: Overcurrent protection and disconnects not shown.

the hydro system finally worked, I decided to double the size of our house, and added four BP 75 watt solar-electric panels and later, four more BP 50 watt panels. The setup complements the hydro turbine, producing at about 420 watts in full sun. I heat my domestic hot water and regulate the output of both my hydro and PV charging systems with a DC water-heating element controlled by a Trace C35 controller.

In the power room, hydro and solar power converge. When the batteries are full, excess hydro or solar electricity is used to heat domestic water.

Independent Life

My son who lives in the city once said to me, "If you hopped, skipped and jumped all the way to town, you still wouldn't get back in step with the world." He and my daughter, who lives a simple life in the country, have been a big part of my inspiration to live up to one of my favorite slogans, "Dare to be different — you might impact the world." When they were young, we took trips to help people in Haiti, and seeing a poor country had a huge effect on all of us. It made us appreciate what we have here and urged me on to be more self-reliant and less wasteful.

I had a dream of an independent life, totally off-grid, trying something that most people would say, "Oh that's impossible. You can't do that." This is what makes me push on. With direction and encouragement, I was driven to build my dream, one building block at a time.

I am working to get young people to view my home and systems, to let them know it's all right to be different, to follow a different path. Maybe it will make an impact and help make the world a better place to live. I might have given up if hadn't received such great advice from Don Harris, and I'd still be in the dark. Thanks also to *Home Power*, hydro prevails![1]

Editor's note: This article first appeared in *Home Power* Issue #108 (August/September 2005) and is reprinted with permission of the author.

Remote Power
and Amateur Radio

PETER TALBOT, VE7CVJ

A recent issue of *Home Power* contained an article about the combination of amateur radio and solar power as a great example of a synergistic blend of technologies. Here's a tale of such a union.

The Need for Power

It all started back in 1979, when my good friend Gordon bothered me enough into studying for my ham radio license. At that time, I had a unique job working in the remote wilds of British Columbia as a ranger at a marine park. I was far from any of the everyday conveniences that we usually take for granted. This was an ideal spot for an independent electrical source and wireless communication, if ever there was one!

So there I was, staying up late at night studying the theory book by candle light and wondering why we still had to know all about vacuum tubes. In the back of my mind was the consuming question: how was I going to power a radio transmitter and the other electrical loads that I could use?

I soon got my license, and before long I fired up my first old radio transmitter, gasping at the 10 amps it drew from the 12 volt battery that I had pulled from my boat. It soon became obvious that there was a need for a steady supply of DC power to run all of the loads that I was adding. As a kid, I made a crude Pelton wheel out of a tin can and pieces of Meccano (like an Erector Set) and used the garden hose to power it. The wheel was then coupled to a bicycle generator. I could run a few lights and an AM radio with it. I thought to myself, "Well, why not do the same thing here, but on a larger scale?"

Amateur Microhydro

There was a stream, which flowed over a sloping waterfall close by. It looked like it could spin some form of wheel and do useful work. Knowing next to nothing at the time about the design of microhydro plants, I set to work with a few lengths of three-inch plastic pipe, plywood and some glue. Before long, I had a 14 inch diameter wheel, mounted on a one-inch steel shaft, and an 80 foot pipeline. The wheel was crude, to say the least. It was more like a paddle wheel than a skillfully crafted Pelton runner. Nevertheless, when I aimed a one-inch jet of water at it with 30 feet of head, the wheel took off!

I scrounged an old Delco alternator and coupled it with a fan belt to a ten-inch pulley on the wheel. I turned on the water and was ecstatic when a meter showed that eight amps were being produced from this simple setup! I ran the output and a length of bare #8 copper wire through the trees 200 feet to the cabin. This kept the battery fully charged all season, and I had plenty of power for lighting and the radio transmitter. On a quiet night when the stream was low, I could hear the soft whirring of that old wheel and the muffled cracks as air under pressure shot out of the nozzle. That first setup was a classic and thrilled me more than anything has since!

Soon, I was able to talk halfway around the world from my small cabin in the woods. I was amazed at how the power of falling water could spin a simple turbine to produce mechanical energy that in turn was being converted into

Peter's 80 meter shortwave (left), PV panel and marine VHF (right)

Peter Talbot

electricity. In the trees overhead, a simple wire antenna was radiating high-frequency radio waves that could be picked up thousands of miles away. I logged conversations to Japan, Australia and even had a contact to Russia.

Practical Aspects

The water-powered ham radio station was also useful in other aspects. I could pass on information for boaters who were making their way up this part of the coast. I did this by contacting other ham operators who would then pass traffic on to family or friends via telephone.

From a safety standpoint, the hydro plant allowed me to call out for help should the need arise. This situation was not long in coming. A man had climbed a steep waterfall and had fallen with rather bad consequences requiring air evacuation and police assistance. The fastest way to get help was to fire up the transmitter, break in on a local conversation and have a phone patch made to the authorities. Within an hour, a plane was tied up at the dock.

Solar Adventures

I was off on a solo kayak trip 400 miles from home. The ultimate destination was a place called Brooks Peninsula on the western coast of Vancouver Island. Brooks is just about as far as one can go on the island as it's a three-day paddle from the road's end. It's the kind of place that demands decent weather and keeping your wits about you. One can never guarantee that there will be other people in such a remote location, so I consider the ham gear and a solar panel essential. I probably wouldn't go to such a place if I didn't have it and couldn't rely on it working when required.

This time, I had a small low-power shortwave radio transmitter on board. I designed and built the transmitter with reliability, small size and light weight in mind for use on backpacking trips and other adventures. The charging source for both the miniature ham radio and a VHF marine radio is a piece of an amorphous solar panel that I cut from a larger broken panel. I mounted the 4x12 inch strip on one-inch thick waterproof foam and then encased it in an aluminum shell with a Plexiglas window. The result is a rugged, fully sea-waterproof and buoyant solar charger. I fasten it on the bow of my kayak so that it gets maximum exposure to the sun. It outputs 14 volts open circuit and 75 milliamps short circuit. The four NiCad cells see about 50 milliamps in average conditions, which is just about right. With the solar charger hooked up, I have never been without the use of the radios even after a long trip.

With the Brooks Peninsula destination in mind, it takes all day to reach the launch point. Much of the afternoon is spent driving over rough logging roads. If I've planned well, I can get the kayak loaded and be under way within half an hour. The concern then turns to navigation along the rocky coastline, watching for bad weather and finding a suitable site for a campsite. Often this takes place in the gathering gloom, illuminated by a small halogen searchlight. A gel cell powers the light, which is solar charged, of course!

When I arrived on the island, I found the perfect place to camp — a small sheltered lagoon with a smooth sandy beach surrounded by dense forest. I arrived early enough to get all the gear set up. Soon, I had the BC Public Service Net tuned in, and was receiving signals from all over the West Coast.

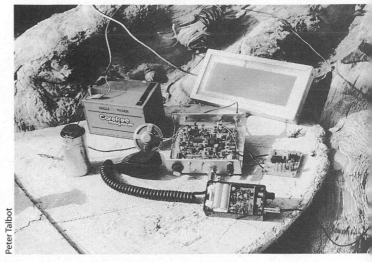

Peter Talbot

Solar, single sideband radio and CW radio (black case)

I generally try to make contact each night on the Net to let the folks back home know of my progress and outline my plans for the next day. It's a comforting thought knowing that if anything went drastically wrong, people would know where I was. This is in no way detracts from the wilderness experience. I enjoy both aspects of the adventure, and it makes good sense to play it safe in these remote locations.

Ham Radio to the Rescue

While collecting driftwood and attending to matters at hand, I was so preoccupied that I scarcely noticed the approaching water. At this time of year, the tides peak in the evening and rise right to the top of the beach where I was camped. I heard a spluttering sound, and saw a cloud of steam as my campfire floated away — the tent was about to get flooded! I grabbed the ham rig and fired it up. Earlier, I had heard another marine mobile check into the BC Net. I figured I'd try to raise him to see if he had a tide table. If it was close to high slack, I just might make it without pulling camp. The word came back from a sailboat 150 miles south of my position that the tide was just about to turn at Brooks, so I anxiously waited. Sure enough, within five minutes the water started to recede!

For the next few days, I passed by some of the finest coastal scenery on the planet and en- countered no one else. The pleasure of working with renewable energy and amateur radio comes alive on adventures like this one. The blend of technologies is perfectly suited to these sojourns into the wilderness.[1]

Editor's note: This article first appeared in *Home Power* Issue #66 (August/September 1998) and is reprinted with permission of the author.

22

Been There, Done That

RICHARD PEREZ, N7BCR

When it comes to capturing renewable energies, it's hard to find a homestead that does more than Bob-O and Kathleen's. The Jarschke-Schultze family uses photovoltaics, wind, microhydro, solar-powered irrigation and solar hot water in their Northern California home. If there's a renewable watt-hour of energy to be had, they are on top of it.

A Personal Note

This renewable energy system displays demented attention to detail. A system as complex as this one takes years to evolve. Very few instantly accomplish what you will see here. In order to understand this system's design, you must first meet the people who live with this system — especially Bob-O Schultze, the system's designer and installer.

Been There

I first met Bob-O and Kathleen in 1988. He and a group of readers visited Agate Flat about *Home Power* Issue #5. They were all living on renewable energy and had to check out this new magazine. Karen and I were amazed. They were the first readers to brave our eight-mile-long four-wheel driveway.

These hardy folks lived along the banks of the Salmon River in Siskiyou County, California. They were a collection of loggers, treeplanters, gold miners, back-to-the-landers and refugees from the cultural wars of the 1960s.

I became fast friends with Bob-O. He and I shared common interests in renewable energy, electronics and radio. Bob-O, Kathleen and Bob-O's son Allen were living beside the Salmon River on a mining claim aptly named *Starveout* due to the seasonal nature of the water runoff needed to mine.

Done That

Starveout was powered by a small hydroelectric system that Bob-O installed in 1980. One of the reasons he came to visit us was to thank me for publishing the Mark VI Field Controller circuit (see *Home Power* Issue #2) which he built to ride herd on his hydro alternator. In 1987, Bob-O and Carl Eichenhofer began manufacturing and selling small hydroelectric turbines called Lil Otto. Bob-O was busy helping electrify the Salmon and Klamath River dwellers with renewable energy and installed over 20 systems along the rivers in five years. But most of the family's livelihood came from

Bob-O and Kathleen's Appliances

Inverter powered appliances	% of total
1 Bob-O's Sony 1730 Monitor	13.8
1 Katheen's NEC4FG Monitor	10.4
1 Bob-O's Mac II Computer	9.8
1 Kathleen's Mac II	8.9
2 Answering Machines	7.4
1 Television Set, Signal Amp	5.3
3 Fluorescent Lights (House)	4.1
1 Fax Machine	3.7
2 Kathleen's Desk Lights	2.6
1 Washing Machine	1.2
2 DeskWriter Printer (idle)	1.1
1 Shoplight	1.0
2 Incandescent Lights	0.9
1 Kirby Vacuum Cleaner	0.9
1 Video Cassette Recorder	0.8
1 Power Tool	0.7
1 Main Stereo System	0.6
1 Microwave Oven	0.5
1 Shoplight	0.2
2 DeskWriter (printing)	0.2
1 Allen's Stereo	0.2
1 B & K's Radio	0.1
1 Makita Battery Charger	0.1
1 Food Processor	0.1
1 Blender	0.1
1 KitchenAid Mixer	0.1
1 Modem Supra V.32bis	10.1
1 Scanner HP IIp (idle)	0.0

Total inverter watt-hours per day (kilowatt-hours per month): 4,848 (145.4)

12 vdc powered appliances	% of total
1 Sun Frost RF-16 Frig/Freezer	9.2
1 Inverter Standby	5.9
1 2 Meter Radio RX	2.2
1 2 Meter Radio TX	0.3
1 2 Meter Radio Amplifier	0.3
1 HF Radio	0.1
1 Metering — CE+, Equus	0.0
1 Soldering Iron	0.0

Total 12 volt dc watt-hours per day (kilowatt-hours per month): 1,179 (35.37)

Total power use in watt-hours per day (kilowatt-hours per month): 6,027 (180.77)

Richard Perez

working the woods — brushing, treeplanting and logging.

In 1990, Bob-O had an accident — a tree he was felling kicked back and crushed his leg. After two weeks in the hospital, he was looking for a new job. With a leg full of metal, logging was out. Kathleen gave him the word, "You weren't fast enough to get out of the way last time, you're a lot slower now." Then, the US Forest Service began cracking down on old mining claims along the Salmon. Starveout, the Schultze's home, was on the hit list. Now Bob-O and Kathleen are serious folks. Rather than wait for the shoe to fall, they listened when Fate spoke. No job, no home. Well, it must be time to move!

And move they did. Bob-O took over Electron Connection, got his California Electrical Contractor's license and began devoting full-time attention to renewable energy systems. Kathleen came to work with us at *Home Power* magazine. They live six miles from *Home Power* Central and two miles from the end of the power lines. Bob-O uses his home as a test bed for new products and system design ideas. Over the years, I have watched their system grow into its present state.

Energy Requirements

Bob-O operates Electron Connection from his home. This means that his computer system is running much of the day to handle the routine business of designing and selling renewable energy systems. Kathleen also has an office in her home with her own computer system. Their renewable energy system supports two full-time business computer systems in addition to their family's domestic power use. The table here details their electric power use.

Renewable Energy Resources

The Schultzes are one of the fortunate few who live at a site that has solar, wind and hydro resources. Bob-O, Kathleen and Allen live next to Camp Creek about seven miles south of the summit of Soda Mountain. A narrow steep valley follows Camp Creek's watercourse and ends at the man-made Iron Gate Lake. From the summit of Soda Mountain to Iron Gate Lake, the land falls over 4,000 feet in less than nine miles. The Camp Creek canyon is a natural wind tunnel driven by cooler air on the mountain and the large lake acting as a thermal flywheel. Water flow in Camp Creek is high during all but the depths of summer.

The most interesting aspect of this site's resource survey is that not one of these sources is reliable enough to provide continuous power. During the winter, the nearby lake provides healthy doses of dense fog and low clouds. During midsummer, the creek slows to a trickle. The wind is strong whenever a weather front passes through or whenever the weather is driving Camp Creek's wind tunnel. It's a case of using what Mother Nature offers when she offers it.

Bob-O didn't start out by capturing all these renewable resources at once. First he developed the photovoltaic system, then the hydroelectric turbine and finally the wind electric generator. It took over four years to build what you see here.

System Design

Bob-O was farsighted when he began designing his system. As the system grew to accept all three renewable energy inputs, only one major change required backtracking—the conversion of the system's battery voltage from 12 to 24 volts DC. This conversion was complex enough that Bob-O has written an article (see *Home Power* Issue # 41) about the process.

The equipment used in Bob-O's system reads like a list of "Things that Work!" product tests. He wants the best and most cost-effective equipment in his customer's systems as well as his own. He refuses to sell a product that he "hasn't tried to break." And being a dealer means that he is exposed to all types of hardware applied in many different systems. Installing dealers, like the ones near you, quickly find out what works and what doesn't.

Renewable energy saves Bob-O and Kathleen US$70,760

PV Electric System

The photovoltaic array consists of 12 Kyocera 51 watt PV modules mounted on a Wattsun two-axis, active tracker. This array produces 18 amperes of current at 30 VDC. With the added assist of the Wattsun tracker, the array produces about 4,000 watt-hours of power on an average sunny day. One hundred and fifty feet (round trip wire length) of #1/0 AWG copper cable feeds the array's power to the house (see *Home Power* Issue #25, page 56 for a "Things that Work!" review of the Wattsun tracker).

Hydroelectric System

Bob-O uses an Energy Systems and Design (ES&D) Turgo-type hydroelectric turbine. Even though Bob-O manufactures the Lil Otto turbine, he uses the ES&D model because it is more suited to his hydro site. A 3 to 2.5 inch diameter, 800 foot pipe snakes its way up Camp Creek. The 27 feet of head created by this pipe supplies the turbine with 9.25 pounds per square inch of working pressure and a flow of 35 gallons per minute. The hydro turbine produces two amperes at 26 VDC or about 50 watts of power. While this may not sound like much power, remember that the hydro is producing 24 hours a day. During a day's time, this hydro produces over 1,200 watt-hours of energy. The hydro's electricity is delivered, unregulated, to the battery via 180 feet (round trip) of #6 AWG cable.

Wind Electric System

This spring Bob-O added a Whisper 1000 wind generator to the system. This wind genny sits atop a 63 foot tower made from 2.5 inch diameter, Schedule 40 steel pipe. The guyed tower is located in a field about 200 feet northeast of the house. This generator produces over 30 amperes at 28 volts in 20 mile per hour winds. Bob-O figures that the wind generator has been producing an average of 2,000 watt-hours of energy per day when the wind blows. Power is transmitted from the wind generator to the house by 380 feet (round trip) of #1/0 AWG cable.

Engine/Generator

Bob-O comes from the group of RE users that would rather eat a bug than start the generator. Nevertheless, Bob-O had to fall back on his 3.5 kilowatt Miller Roughneck generator/welder several times last winter (before the Whisper 1000 was up and running). He hopes the addition of the wind generator will permanently retire the Miller from generator service.

Batteries

This system uses eight Trojan L-16 lead-acid batteries to store energy. Each L-16 battery is rated at 350 ampere-hours at six volts DC. The battery is configured at 700 ampere-hours at 24 VDC. Each cell in the battery is fitted with a Hydrocap® which recombines gaseous hydrogen and oxygen into pure water. These Hydrocaps not only keep the system safer by nearly eliminating the potentially explosive hydrogen gas, but reduce cell watering and battery top cleaning. The battery is located in the home's basement along with the inverter and power processing gear. The battery interconnect cables are made from #00 AWG copper cable with soldered ring terminal ends. All the batteries are sitting in Rubbermaid™ plastic tubs just in case there is any spillage of electrolyte.

Inverters

One of the major reasons that Bob-O converted the system from 12 to 24 VDC was to accommodate the new Trace 4,000 watt sine wave inverter. The inverter converts the low voltage power stored into the battery into 120 VAC, 60 hertz sine wave power like the utility rents out. This new Trace inverter has been performing faultlessly since installed four months ago. Over the years, Bob-O has used just about every inverter available, and he thinks the new Trace is a definite "keeper." The inverter's output is wired directly into the home's main panel where it is distributed to all the home's branch circuits. Since the inverter produces sine wave power, all of the appliances in the house perform just like they were plugged into the utility.

Regulators

Bob-O uses a Heliotrope CC-60B PV controller (see *Home Power* Issue #8, page 31) set to regulate at 31 VDC. This is a little high, but the business uses so much power that Bob-O feels he'll take an equalizing charge whenever he can get it. The hydroelectric turbine produces less than 100 watts and is not regulated. At this point in time, the Whisper wind generator is also not regulated. This has led to several inverter shutoffs from battery overvoltage. Bob-O's next project is getting the load diversion feature of the new Trace inverter to dump his excess power into heating water in the 80 gallon DHW tank. Once this is accomplished, the Whisper will be effectively controlled and all the system's surplus power will be diverted into making hot water.

Converters

When the system changed from a 12 volt battery to a 24 volt battery, Bob-O was faced with a decade's worth of 12 VDC appliances. Most were replaced by 120 VAC models, but several stubbornly remained 12 volt. In order to power this 12 volt gear (like a Sun Frost RF-16 refrigerator/freezer and a whole rack of 12 volt ham radio gear), Bob-O uses a Vanner Voltmaster. From a system design standpoint, the Vanner Voltmaster is a switching power supply that can efficiently convert power stored in a 24 VDC battery into 12 VDC for appliances.[1]

Instruments

Bob-O is an electronics nerd, and his home is festooned with instruments of all types. Only two are in daily use to assess the system's performance—a Cruising Amp-hr+ meter and a homemade, expanded-scale battery voltmeter. The Cruising Amp-hr+ is a battery ampere-hour meter that functions like a gas gauge for batteries. In addition to calculating ampere-hours in and out of the battery, the meter also measures battery current and battery voltage (see *Home Power* Issue #26, page 59 for a review of this Cruising meter). The analog expanded-scale battery voltmeter is a very simple homebrew project (see *Home Power* Issue #35, page 92 for a schematic of this analog battery voltmeter).

Water Systems

The main water source is a spring located about 200 feet in elevation above the house. This spring provides gravity flow water for the house, but hasn't sufficient flow to supply Kathleen's many gardens. Bob-O uses a PV array direct water pumping system to supply over 1,500 gallons daily to the gardens. This system uses two Kyocera K51 PV modules powering a 24 VDC Flowlight Slow Pump. The PVs are mounted on a one-axis Zomeworks tracker, and their power is processed by a Sun Selector LCB before being sent to the pump. This system is simple, effective and uses no battery. The water is pumped from Camp Creek into two 1,350 gallon water tanks located about 40 feet in elevation above the gardens.

Bob-O uses a rack of 20 Thermomax evacuated tube, heat pipe, solar collectors to heat water for the house. This system has been operating for over two years and has survived numerous hard freezes and inch-sized hailstones. These evacuated tubes have the insulation value of a vacuum bottle. Inside each 2½ inch diameter glass tube there is a finned heat pipe partially filled with an alcohol/water mixture. Sunshine causes this mixture to boil, and heat is transmitted to a glycol mixture which in turn transfers the heat to the home's 80 gallon Rheem SolarAid hot water tank. This DHW system is rather complex with two stages of heat exchange and single Laing pump (driven by 0.25 amps at 12 VDC). The reasons to undergo this degree of complexity are absolute freeze-proofing and the incredible cold/cloudy weather performance of the Thermomax collector. On sunny winter days when the ambient temperature is well below freezing and the wind is blowing, the Thermomax still delivers 180° F to the hot water tank. Bob-O also has a Myson on-demand, propane-fired water heater on line. This Myson has the happy ability to moderate its heat output in relation to the incoming water's temperature. If the weather has been sunny and the solar hot water heater has been producing, then the water passes straight through the Myson without any additional heating. Using the on-demand heater as a last resort ensures that the house will always have plenty of hot water regardless of the weather or the amount of hot water needed. This hot water system supports two bathrooms, a kitchen sink and a washing machine. Between the months of May and October the pilot light on the Myson is shut off and the hot water needs are met by the Thermomax alone. Kathleen has a sign above the sink for visitors that reads, "Caution — Solar Heated Water — HOT!"

System Performance

Well, there is never a power outage at Bob-O and Kathleen's place. The photovoltaic array produces about 4,000 watt-hours of power daily. The wind generator is a newcomer to the system, and we don't yet have years of data on its performance. If the wind is blowing, then Bob-O reports that the Whisper makes about 2,000 watt-hours of energy daily. The small hydroelectric turbine produces about 1,200 watt-hours of energy daily. Bob-O figures that he puts about 25 hours of operating time on the Miller engine/generator yearly. This system is about two-thirds powered by photovoltaics, with the remaining one-third divided by wind and microhydro.

The battery in Bob-O's system contains enough energy to power their homestead for about three days with no RE power input whatsoever. And since every day contains at least some renewable energy, the battery is virtually never fully discharged.

System Cost

The tables here detail the costs of all the renewable energy equipment. Bob-O and Kathleen have invested just about US$20,000 in their electric renewable energy systems. While this sounds like a lot of money for power, let's examine the alternative.

Bob-O and Kathleen's property is located 1.7 miles from the end of the utility's power lines. The local utility, Pacific Power, charges US$10.35 per foot for new line extensions. The going local rate for electric power is US$0.095 per kilowatt-hour. Bob-O and Kathleen consume an average of about six kilowatt-hours daily. Figures 3 and 4 compare the cost of running in the utility lines versus using renewable energy. These tables do make some assumptions. One is that the renewable energy system lasts ten years, which is far more certain than the second assumption, that the utility will

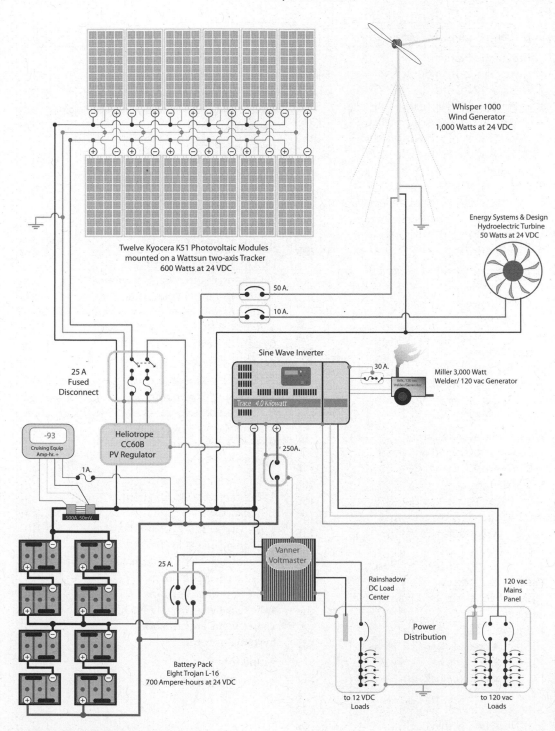

Whisper 1000
Wind Generator
1,000 Watts at 24 VDC

Twelve Kyocera K51 Photovoltaic Modules
mounted on a Wattsun two-axis Tracker
600 Watts at 24 VDC

Energy Systems & Design
Hydroelectric Turbine
50 Watts at 24 VDC

50 A.

10 A.

25 A
Fused
Disconnect

Sine Wave Inverter

30 A.

Miller 3,000 Watt
Welder/ 120 vac Generator

Trace *4.0 kilowatt*

-93
Cruising Equip
Amp-hr. +

Heliotrope
CC60B
PV Regulator

1A.

250A.

500A. 50mV.

Vanner
Voltmaster

25 A.

Rainshadow
DC Load
Center

120 vac
Mains
Panel

Power
Distribution

Battery Pack
Eight Trojan L-16
700 Ampere-hours at 24 VDC

to 12 VDC
Loads

to 120 vac
Loads

Home Power

Bob-O and Kathleen's Systems

Bob-O and Kathleen's System Cost

Photovoltaic System

12 Kyocera K51 PV modules	$4,200
1 Wattsun 12 PV dual axis tracker	$1,575
1 Heliotrope CC-60B charge controller	$295
1 C&H 60 ADC fused safety switch	$215
5" × 10' steel pipe, cement, gravel	$150
150' 1/0 AWG THHN main feeder wire, 1¼"	$137
PVC conduit, NEMA3J box	$70
84' 10 AWG USE, PV interconnect wire,	$27
Crimp wire terminals, split bolts, tape	$25
1 8' Copper ground rod, clamp, wire	$15
System Subtotal	**$6,708**

Hydroelectric System

1 ES&D FT1 hydro with 24 volt low head stator	$830
600' 2½" PVC 160 pipe	$420
200' 3" PVC 160 pipe	$244
Valves, fittings	$60
90' 6 AWG triplex wire	$45
1 SquareD QOCB box with DC circuit breaker	$42
System Subtotal	**$1,641**

Wind Generator System

1 Whisper 1000 wind generator	$1,500
380' Wire 1/0 THHN	$346
8 2½" flanges	$208
105' 2½" Sch 40 steel pipe	$160
700' ¼" aircraft cable	$158
1 1¼" PVC conduit, NEMA3 JBox	$135
Sand and gravel	$130
8 ⅝" × 12" turnbuckles (surplus)	$96
Miscellaneous steel	$50
Miscellaneous wire, terminals	$50
Cement	$42
1 SquareD QOCB box with DC circuit breaker	$42
48 ¼" cable clamps	$29
12 ⅝" bolt with nylock nut	$11
20 ¼" thimbles	$11
2 ¾" × 6" bolt with nylock nut	$7
6 5⁄16" × 5" bolt with nylock nut	$2
System Subtotal	**$2,976**

Batteries

8 Trojan L-16 lead acid batteries	$1,440
24 Hydrocaps™	$180
11 2/0 AWG, 13.5" battery interconnects	$107
System Subtotal	**$1,727**

Inverter

1 Trace SW4024 with conduit box	$3,045
1 Heinemann 250 amp breaker with enclosure	$245
2 Trace BC-5 4/0 inverter cables	$150
1 2" PVC conduit, fittings	$12
System Subtotal	**$3,452**

DC Load Center, Metering

1 Cruising Equipment Amp Hour+™	$325
1 20 amp Vanner Voltmaster	$304
1 Rainshadow DC load center with 4 CBs	$215
1 SquareD QOCB xox with DC CBC	$52
System Subtotal	**$896**

Solar Irrigation System

2 Kyocera K-51 PV modules	$700
1 Flowlight® Slowpump	$488
1 Zomeworks two-panel TrackRack™	$385
1 Sun Selector LCB model 3MT	$80
Wire, fused disconnect	$75
System Subtotal	**$1,728**

Solar Hot Water System

1 Thermomax SOL 20S thermal collector	$1,723
1 Myson CF-325-2 demand heater	$610
1 Rheem SolarAide 80 gallon tank	$525
1 Heliotrope Delta T thermostat/ control	$140
1 Laing circulation pump	$125
1 Amtrol expansion tank	$50
1 Relief valve - Watts 174A	$44
Valves, vents and sensors	$120
System Subtotal	**$3,337**
Grand total	**$22,465**

Note: All figures are in US dollars.

Richard Perez

Utility Versus Renewable Energy

Access	Utility Power Cost	Renewable Energy Cost
Line extension cost	$88,762	
RE system cost		$19,207
Ten year power bill	$2,081	$0
Maintenance	$0	$875
Ten year cost	$90,842	$20,082
$ per kilowatt-hour over ten years	$4.15	$0.92

Renewable Energy saves Bob-O and Kathleen US$70,760

Note: All figures are in US dollars.

(Energy consumption = 6 kilowatt-hours per day

(180 kilowatt-hours per month)

Distance from utility lines = 1.7 miles

Richard Perez

not raise its power cost in the next ten years. I figure that Bob-O and Kathleen saved more than US$70,000 by using renewable energy for electricity.

If you consider that a new truck costs about US$20,000, it's easier to understand Bob-O and Kathleen's investment in self-sufficiency and clean energy. In terms of performance for money spent, I pick an RE system over a gas guzzler any day.

Being Here Now

Bob-O and Kathleen live on an energy self-sufficient homestead. Their dedication to a sustainable future that all can share makes them friends of all living on this planet. I salute them![2]

Editor's note: This article first appeared in *Home Power* Issue #41 (June/July 1994) and is reprinted with permission of the author.

The Ten Kinzel/Kingsley Rules
for Surviving Microhydro

TERRY KINZEL AND SUE ELLEN KINGSLEY

In the fall of 1991 we started to build a small, off-the-grid house. Living next to Lake Superior in Michigan's upper peninsula, we knew that a source of electricity to supplement our PV panels would be necessary to get us through our dark and cloudy winter. The tall towers required for wind turbines were quite daunting. A stream flows through our yard, but thinking it viable only out west where the heads were high, we didn't seriously consider hydro power initially.

Attending the Midwest Renewable Energy Fair prompted us to reconsider hydro power and take actual measurements. We consulted with Paul Cunningham of Energy Systems and Design. Now, while our PV panels are an idle piece of art during the long night of December, our hydroelectric turbine produces generous, reasonably reliable power. Now in our third year of hydro power, this satisfactory state has not come without glitches. What follows is a Murphy's Law catalogue of things that will go wrong for any ordinary person attempting to grapple with microhydroelectric power.

Rule Number 1

Never underestimate your ability to cheat on your measurements.

We measured the flow of our stream using two methods (see *Home Power* Issue #8, page 17 and *Home Power* Issue #15, page 17) coming up with about 750 gallons per minute in the driest month. Since this was so much more than we needed, an overestimate wouldn't have caused much of a problem. Measuring head was another story. We used a 50 foot garden hose, stretched out in the stream bed. After a flow through the hose was established, we would raise the downstream end. The distance from the stream surface to the hose end was then measured, giving an estimate of the head over that section of the stream. The process was repeated until the portion of the stream from the proposed intake to the turbine was measured. Errors are easy to come by. There is at least a two-inch difference between where the flow just begins to stop and where it actually quits. Inertia tends to accentuate this error. The stream surface is usually rippled. Of course, we erred on the side of more apparent head. We were off two feet over the 400 feet of the stream bed. We thought we had 17 feet of head when the reality was 15 foot of head.

This error was compounded by minimizing the height above the stream bed that the turbine must be placed so as not to be endangered by fluctuating water levels — allowing us to pretend we had a foot of head more than we actually did.

Rule Number 2

Never underestimate the ability of the technical elite to dazzle and befuddle us technological dummies.

Rule Number 2A

Never underestimate the ability of the technical elite to overestimate the knowledge of us technical dummies or to take for granted critical issues which seem obvious to them because they work with them daily, but are anything but obvious to us.

Having decided that our site had potential, we called Paul Cunningham at Energy Systems and Design, who after hearing of our site said something to the effect, "Whoa, you'll have so much electricity that it will be too cheap to meter." (Reminding me of the infamous promise of atomic power.) He subsequently launched

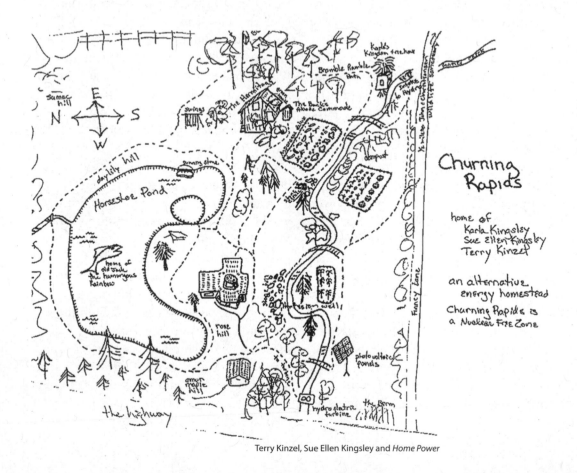

Terry Kinzel, Sue Ellen Kingsley and *Home Power*

into about 500 calculations in the next few moments, occasionally asking a question in some language faintly reminiscent of English. Having only the vaguest idea of the meaning of the questions and not wanting to appear too foolish, we gave answers we hoped would please him. The upshot: a shiny new turbine with the cutest little runner (water wheel) appeared in our garage a few weeks later.

Rule Number 3

Never underestimate friction.

Our site (using our somewhat inflated values for head) called for a FAT (Ford Alternator Turbine). The turbine used two ¾ inch diameter nozzles, delivering 75 gallons per minute and yielding a predicted output of 90 watts. Given the projected run of 300 feet (underestimated from the actual 350 feet — see Rule #1), 17 foot head and 75 gallons per minute flow, a four-inch drainpipe was chosen to deliver the water. This was split just before the turbine and stepped down to two 1.5 inch pipes attached to the two nozzles. The turbine was bolted to a cement block sitting about two feet above the stream with a six-inch stove pipe running through it to handle the water egressing from the turbine. Since there was only a single four-inch pipe delivering water, it seemed obvious that a six-inch pipe for the tailwater would be sufficient. After some minor missteps, the water was hooked up and the wires connected in approximately the correct order. We let her rip, anticipating the glorious vision of the ammeter plunging off the scale as power surged through our circuits.

In fact, it was hard to tell that the analog ammeter in the turbine moved at all. However, using a digital ammeter and voltmeter,

it appeared that the turbine was producing 1.2 amperes at 13.0 volts or a dazzling 15.6 watts. Despair! Grief! Frantic calls! "Describe your setup again." "What about egress?" "You need absolutely free egress so that friction won't slow down the wheel!" (See Rule #2A.) So we modified the system to lower the turbine to about one foot above the stream and gave it free egress, bringing the output up to 45 watts. The turbine proved very reliable, causing no problems throughout the winter. The below-predicted output was not a problem since there was still construction going on, and we were still partly connected to the electric utility grid.

Rule Number 4

Never underestimate the capacity of technical dummies to learn and be helped by considerate, one-on-one, face-to-face consultation.

The next summer after having a year of generally positive experience with our turbine, we returned to the Midwest Renewable Energy Fair. We were still vaguely unhappy that it was producing only about half the predicted power. After a delightful consultation with Don Harris, we made the following modifications:

1. Increased the nozzle size to two one-inch nozzles (included with the original order).
2. Laid a second four-inch pipe to decrease friction loss in the pipe, especially with the increased flow through the larger nozzles (see Rule #3).
3. Replaced a section of line that had been squished a bit by a gravel truck driving over it (see Rule #3).
4. Lowered the turbine to six inches above the stream bed.

5. Raised the dam at the intake site about eight inches. A board across the stream creates a pool deep enough to cover the screened intake.

The result: 115 watts of continuous power (2.75 kilowatt-hours daily) for the past 18 months (with a few dramatic interruptions).

Rule Number 5

Never underestimate the power of water to wreak havoc.

A four-inch pipe under only seven or eight pounds of water pressure doesn't sound very threatening. However, when a joint pops free at the last fitting before the turbine with you being the only object between an ocean of pressurized ice water and an expensive piece of electrical equipment, the experience is distressing.

Our intake is screened by hardware cloth and window screen in a wood frame into which the two four-inch pipes fit. During most of the year, it requires no attention. However, during the spring melt and the fall leaf season, it periodically needs to be cleaned. A grass rake handles this task. However, during the times the screen is occluded with leaves, the columns of water in the pipes create a huge suction. On more than one occasion, the suction has collapsed the box or sucked the screen into the pipe. Always build this part of the system stronger than you ever dreamed necessary (see Rule #2A).

It rains, the stream surges and the dam you thought was stronger than Grand Cooley washes out—a scenario guaranteed at least once. Although inconvenient, this allows opportunity to fulfill every little boy's dream of playing in streams.

A corollary here is: Don't Get Greedy. After coming to fully appreciate the importance of head and pressure, we tried to squeeze the most power possible out of the turbine. We moved it as low above the stream as seemed safe. The same "once in a decade" fall storm that washed out the dam caused the stream to surge within millimeters of the turbine. Being away for the night (Rule #8), we only realized this later. Fortunately, during the winter when we need the most power from the turbine, the stream is very steady. During the other seasons, the PVs produce so much power that the turbine can be raised safely out of harm's way.

Rule Number 6

Never forget that even moving water freezes.

Water abhors discipline. The board we installed to raise the intake pool is buried in the stream bed. Water flows over the top of the board. Last winter, when the mercury hit −20°F, the top of the pool froze over, and the water chose to dig a channel underneath the board. The intake was left high and dry. We filled burlap bags with stones to span the breach, and it held for the rest of the winter. No fingers or toes were lost to frostbite.

Having watched the stream for many winters, we knew it never got more than a crust of ice. Since most of the pipe was buried, we were not too worried by the prospect of freezing. The first winter, we lightly insulated the small portion that was exposed. Our actions were somewhat validated when we experienced no freezing problems. We went into the second winter with a modified system. With two four-inch supply pipes, the water flowed more slowly. Also, the turbine nozzles are on opposite sides. So, one pipe is a straight shot while the other is forced to make a 180° loop to reach the back side, slowing the water further and exposing more pipe to subfreezing air. That winter was the coldest in many years. After our third night of 25 below, with highs reaching all of −15°F, we awoke to an output of about 50 watts. Sections of the long and winding pipe were frozen. We were resigned to the idea that the entire pipe would now freeze solid and wouldn't thaw 'till summer. The next two days were above zero, and for reasons that remain completely obscure to us the pipe thawed. We beefed up the insulation in exposed portions and maintained full power for the remainder of the winter. As a consequence of this experience, we modified the pipes last summer. Both delivery pipes each made a 90° turn and were stepped down from four inches to two inches in diameter before the bends. We reasoned that the water would be moving faster through the two-inch pipes and would be less likely to freeze. Unfortunately, this resulted in a 15 watt loss of power (see Rule #3). Consequently we went back to the original design and put a bit more insulation on when the snow began to fly.

Rule Number 7

Never forget that, for most of us, electricity moves in mysterious ways.

· The first year, after we got the output up to 45 watts, we were troubled by the fact that the voltage at the turbine always read about 13.5 to 14.0 volts. This did not seem high enough since our PV panels were producing 17.8 volts, and we were using NiCd batteries (since replaced with lead-acid) with a fairly high voltage. Although we had plenty of power (our 120 VAC circuits were still grid-connected at that time), we weren't quite sure where the electricity was moving. The low voltage was suspect

in the below-predicted output. This was before we really believed Rule #3. Several calls to New Brunswick regarding this matter enriched Bell Telephone and reconfirmed Rules #2 and #2A. We returned the turbine. Paul stated that it worked fine and he couldn't understand why we were upset about the voltage. He managed a rapid turnaround time, paid for return postage and installed a new, more efficient runner — all at no charge. Eventually, we came to realize that the open circuit/no load will be quite high, while the working voltage will always remain about 0.5 volts higher than that of the battery bank. The electricity always flows in the correct direction. Why this is so remains a mystery to us. By the way, why is the sky blue?

Rule Number 8

Never will your hydroelectric system need attention when it is convenient.

This hardly needs elaboration, but be especially vigilant around the times you have purchased expensive, nonrefundable airline tickets — The System Knows.

Rule Number 9

Never will any local contractors, local electricians or your friends know enough about your system to easily solve a problem.

In dealing with a problem, a mechanically-oriented and long-standing friend is your best bet. A corollary to this rule is: tell a house sitter how to read the meters and how to shut the system off when there's trouble.

Rule Number 10

Never is the power output of your hydroelectric system affected by the phase of the moon or your menstrual cycle.

Churning Rapids Fact Sheet

Property: 2.7 acres
House size: approximately 750 square feet
Builders: primary, Brian Maynard; secondary, Dan DePuydt and Dave Bach
Design: Terry Kinzel

Energy Production

Photovoltaics: Eight Solarex MX60 PV modules mounted on a Wattsun tracker producing 480 watts (28 amps at 17.4 volts in full sun)

Hydroelectric: Energy Systems and Design Ford Alternator Turbine with 16 feet of head, flow of about 75 gallons per minute producing 115 watts (9.4 amps at 12.2 volts, continuous)

Energy Storage: Six L-16 lead-acid industrial batteries in series and parallel to give about 1,050 amp-hours at 12 volts

Energy Management: Enermaxer charge controller with two 15 amp hot water resistance coils to preheat water — in sum providing a substantial portion of our hot water

Inverter: Trace 2012 (has trouble with the clothes washer — inquire for details)

Metering: Cruising Equipment amp-hour meter, two SCI Mark III meters measuring battery voltage, amps in from PV, amps in from hydro, amps out through DC junction box and amps out through the inverter

Heat Source: Reliance high efficiency wood stove with Olympic catalytic propane heater backup

Hot Water: Enermaxer preheat, Aquastar instantaneous propane heater

Well: 362 foot artesian well; Flowlight booster pump to pressurize the system

Appliances: Sun Frost 12 cubic foot refrigerator, Sun Frost 10 cubic foot freezer, Caloric propane range and oven, Kenmore front-loading washer

Lights: Electronic ballast compact fluorescent and 12-volt halogen incandescents

The house is a modified superinsulated design (not completely airtight and too much window area for maximum efficiency). All glazing is high performance — mostly Anderson windows.

There is too much plumbing, partly due to the two-part development of Churning Rapids. The showers use low flow head. The toilets are Kohler one-gallon flush connected to a standard septic system and drain field. The Buck's Adobe Commode Composting Outhouse near the garden gets much use.

Other features: The Hermitage is a guest house/retreat which is tucked into the rafters atop the greenhouse and sauna with dressing room. Pond, veggie gardens, flower gardens, bitsy woods and rambling paths.

Terry Kinzel and Sue Ellen Kingsley

Check the output at least daily; it will be monotonously steady. If the power has fallen off even a few watts, don't look to the moon or consult your calendar looking for the reason. Rather, prepare to get wet.

Conclusion

This has been a summary of our experience with our microhydroelectric turbine. Since we tried to describe some of the pitfalls that may be experienced by people of ordinary skills, it may seem that we are negative. This is not the case. While microhydro is not as simple (for the end user) as plugging into the grid, we have, with help, been able to solve each problem. With a modicum of maintenance and troubleshooting, microhydro has provided us with a generous supply of electricity and allows us to live very comfortably disconnected from the grid. Our batteries have it easy. They are never deep cycled. While most of the technical people we've dealt with suffer from the truths of Rules #2 and #2A, the equipment and service we received from Paul Cunningham were excellent. The advice from Don Harris at the Energy Fair proved invaluable. We would not hesitate to work with either of them in the future.[1]

Editor's note: This article first appeared in *Home Power* Issue #47 (June/July 1995) and is reprinted with permission of the authors.

PART III

Low Head Sites

As well as the classic high head sites and those which work on domestic water pressures, there are many systems generating electricity from drops that are too small even for household use. I call these low head sites. The sites described in this part range from 1.9 feet to about 15 feet of head. The upper limits of low head turbines and the lower limits of higher head, impulse turbines such as the Turgo overlap considerably, so that a variety of technologies might be appropriate at a particular site. Since reaction turbines are so different from the more common impulse turbines, as well as for consistency, let's also say that low head sites in this part use reaction turbines like the PowerPal.

Low head seems to invite innovation. This part presents examples which counter many easy generalizations. For example, there are reasoned explanations of why generating electricity with old style waterwheels is difficult, but an overshot waterwheel generates power in Chapter 26. Or although the conventional wisdom is that multiple kilowatt AC systems are unlikely at these low heads, the systems in Chapters 25 and 26 do just that.

Chapter 24 presents an excellent collection of old time, very low head sites using reaction turbines such as the Francis. An old Francis-style turbine powers Chapter 25. More modern reaction turbines are used in Chapters 27 and 28, with an axial flow turbine at work in Chapter 29. Most of the turbines in the ESMAP project in Ecuador (Chapter 30) were low head turbines operating at five feet or so, with a scattering of higher head sites. Technically, there must be many more opportunities for low head sites than for high head sites, despite the preponderance of higher head turbines presently operating.

Wide experience of low head sites is necessary to comprehend the full range of possible solutions at a particular site; now you can tap the collective experience of these authors. Remember, there is no rule that you have to use all the head available at any site. There may be opportunities for low head solutions even when more head is available. Because the low head turbines can use inexpensive alternatives to expensive piping (such as ditches) the low head alternative deserves close consideration.

24

Ultra-Low Head Hydro

CAMERON MACLEOD, N3IBV

One hundred years ago low head hydro wasn't just an alternative; it was the best alternative. Unlike high head sites, low head sites are everywhere — and often closer to population centers where the power is needed. Power sources were valuable and sought after, because cheap power wasn't delivered through silent wires down every street. Local wars were fought over water rights.

The History of Low Head Hydro

Times have changed, but the weight of water and gravity remain the same. Once we had over two hundred makers of small water turbines in the USA. Some of them built, by 1875, equipment that was 80% efficient. They built and inventoried turbines as small as four inches in diameter that made one horsepower on ten feet of head. Turbines that ran on two feet of head and made from one to fifteen horsepower were common. Some were excellent machines that ran with little maintenance for years. The know-how and hardware were everywhere. In the eastern part of the US, the power of the small streams near populated areas was developed and put to work. All the way from the hills to the sea, this water was

used over and over again wherever topography supplied enough head. One large stream in the east had dams and still has pre-revolutionary deeded water rights wherever early settlers found three feet of head.

When ships landed on the east coast of North America, surveyors and mapmakers headed inland to discover natural resources. All the old maps denoted power sites as *Mill Seats* long before settlers arrived. This was before the successful use of stationary steam engines, so we know that they were referring to hydro power. Later, towns grew because of this power. Virtually every sort of agricultural and industrial work was once aided by the water. It is sad that the water source of power is often blamed today for the mess that industry left behind. In this age of environmental awareness, we should not throw out the turbine with the wash water.

Back when power was valuable, people moved hundreds of tons of earth and rocks with just their backs, mules or oxen. Often they made this investment and did this work with their bodies for the sake of one or two horsepower. Wow! Think about it. Something was going on there. If you think they were

A Typical System

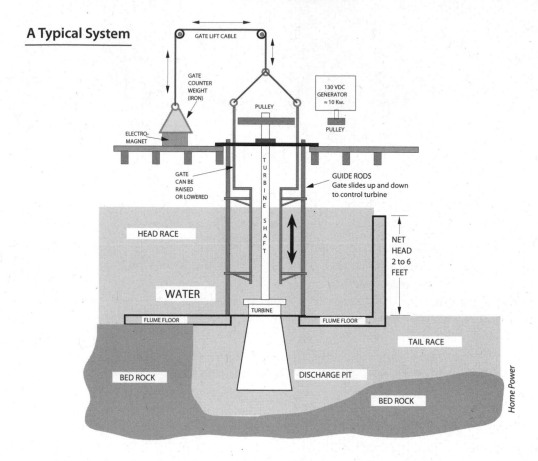

GATE LIFT CABLE

GATE COUNTER WEIGHT (IRON)

130 VDC GENERATOR ≈ 10 Kw.

PULLEY

PULLEY

ELECTRO-MAGNET

GATE CAN BE RAISED OR LOWERED

T U R B I N E S H A F T

GUIDE RODS
Gate slides up and down to control turbine

HEAD RACE

NET HEAD 2 to 6 FEET

WATER

FLUME FLOOR

FLUME FLOOR

TURBINE

TAIL RACE

BED ROCK

DISCHARGE PIT

BED ROCK

Home Power

nuts, then look at the size of the manor houses and mills that were energized with those one or two horsepower. Then think about what clean renewable power in your backyard is really worth to you and your children and your grandchildren — and on and on — forever.

Of course power has gotten cheaper and cheaper in the last hundred years. By burning non-renewable fossil fuels at the expense of the earth and our futures, they practically give it away. I can hear you now — what's this jerk talking about. The only ones that really know the value of power are the people who have tried to make power for themselves. If your goal is to supply your daily energy needs; you either know how cheap commercial power is or you're going to find out. My position is not to discourage you, just to warn you. Pursue your dream. If you can't visualize it, it will never happen.

Over the past ten years, I've helped to develop twenty or so small hydro sites. I've gone on to bigger megawatt hydros now, because I need to make a living. The small sites range in power from 300 watts to 100 kilowatts. Almost all of this work has been under fifteen feet of head. The power has been utilized to run homes and small businesses or more commonly, large farms. All the projects were former sites with dams in one state of repair or other. The legal aspects of these undertakings have been handled by the owners and often represent the greatest problem.

Hydros and Red Tape

If your home power system isn't on federal land, doesn't hook to the grid and doesn't make power from a navigable stream; then you may not need a federal license. There is no legal way to avoid dealing with a state agency. Watch out—often this destroys dreams. You had better base your work on an existing dam or a pile of rocks no more than 36 inches high called a diversion weir. Remember not a dam, but a weir. That diversion had better not be long in either case if you hope to stay within environmental laws. In all cases you had better own both sides of the stream. These problems will vary from state to state. You must learn through research. Have enough sense to keep your own counsel (keep your mouth shut about plans) until you figure out which way the water flows.

Low Head Hydroelectric Turbines

My goal here is to let home power people know that under just the right circumstances low head hydro is possible. Practical—that's your judgment. It will depend a lot on what you consider to be valuable. That is to say, your values. How much your alternatives cost matters too. Despite all this red tape nonsense many people have successfully established low head hydro systems. I'll detail a couple of sites to whet your imagination. First, you should understand that very little has been written about low head hydro in the last 50 years. By 1915,

development had shifted from small diverse sources of power to large centralized systems based on alternating current and high voltage distribution. Giant government-backed utilities were beginning to carve up the country into dependent territories. Starting with the cities and industrial areas they stretched their wires out into the country. By the 1930s, rural electrification was well under way. Many utilities forced their customers to take down their wind machines and remove their turbines before they could hook up. Big customers were bribed with no cost changeovers from DC to AC. Along with the gradual loss of public self-reliance, the end result for the hydro power machinery business was that the market for small turbines disappeared. So did the manufacturers. Several companies made the transition to giant utility-grade equipment into the 1950s. Now they are gone too. None of the biggies are US-owned.

There are a few crazies like myself who still build small machines. Most backyard operations concentrate on Pelton and crossflow turbines which are only suitable for high head (depending on power requirements). I build Francis and propeller-type turbines. They are expensive, hand-built machines that don't benefit from mass production. They will, however, last a lifetime with only bearing changes. This is a tall order because everything must be constructed just right. I approve all site designs before I'll even deliver a turbine. I personally design most systems.

Often a better way to go involves rehabilitating old equipment. Some hydros were junk the day they were built. Other makers really knew their stuff. Their quality and efficiency are tough to match even today. These machines are usually buried under mills or in the banks of streams. Go look, you'll find dozens. The trick is to know which one you want, so do your homework before buying an old turbine.

A Low Head Hydro System

One site that depends on a rehabilitated machine belongs to a farmer named George Washington Zook. George decided not to use commercial power in 1981. He had deeded water rights and the ruin of a dam on his property. Best of all he had lots of water, and incredible determination, common sense and know-how. He only has 36 inches of head. I supplied him with a 30 inch diameter, vertical axis Francis-type turbine. This turbine was built by Trump Manufacturing Co. in Springfield, Ohio around 1910…one of the good ones. George was 25 years old when he finished the project.

George got all the required permits and built a 60 foot long, 36 inch high, log dam with a wooden open flume for the turbine at one end. He installed the turbine with a generator mounted on a tower to keep it dry in high water (never underestimate high water). Four months later his dam washed out. One year later he rebuilt and started generating 130 volt DC power. Yes, high voltage DC. His machine develops 35 amps at 130 volts or 840 ampere-hours per day or 109.2 kilowatt-hours per day. Discharge is 2,358 cubic feet per minute (lots of water) at 96 revolutions per minute. He has a 90 series cell, 240 amp-hour NiCad battery pack. This represents an incredible amount of power for any home power system. That is 32,760 kilowatt-hours a month. Hey, that's enough power to run three to five av-

erage North American homes. All of this on 36 inches of head. Yeah, that's right, and his battery pack lets him meet 20 kilowatt peaks. Here is what his load looks like:

- three freezers (two for the neighbors), a refrigerator, refrigeration to keep the milk from 20 cows cold
- a vacuum system to milk these cows
- two hot water heaters, all lighting in home, barn and two shops
- occasional use of silage chopper, wringer washer, water pump, iron and farm workshop machines

I'm afraid it still goes on — his nephews put in a complete commercial cabinet shop two years ago. They have all the associated equipment including a 24 inch planer. Well, now what do you think about low head hydro?

There are a few key differences between George's system and most you read about. There isn't an inverter on the property. At 120 volts DC, line losses are at a minimum. We have some 220 volt, three-wire systems operating. All of the equipment and machinery on the farm was converted to 120 volt DC motors, including refrigeration. The high efficiency of this approach makes all the difference.

AC Versus DC Hydros

Stand-alone AC is a possibility, but it requires a larger turbine and more year-round water to meet peak loads. The cost of an electronic load governor and the inefficiency of single phase induction motors are two of the drawbacks to consider. Backup generator cost is also a factor. You'll need a big one to meet AC peak loads. With batteries to meet peak, a small generator will suffice.

Remember, if you can meet 20 kilowatt peak loads with batteries it only takes one horsepower 24 hours a day to run the average American home. This is a tiny turbine that uses little water when compared to the 40 horsepower turbine on the same head that would be needed to meet the same peaks on conventional AC. Forget it — there is no comparison. The big machine would cost a fortune and require massive amounts of water. Hey, it is possible, I've built them.

The best of both worlds would have the lighting and heavy motor loads on 120 volt DC for efficiency. It would have a switching power supply running on 120 volts DC putting out high-current 12 or 24 volts DC to run an inverter for specialized AC loads like TVs and stereo systems.

Some Low Head Hydro System Specs

Here are the pertinent details on some stand-alone, DC, low head hydro sites that I've been involved with:

System 1

Five feet of head — eight-inch MacLeod-built CMC vertical Francis-type turbine develops three amps at 130 volts or 72 amp-hours per day or 9.36 kilowatt-hours per day. Discharge is 72 cubic feet of water per minute at 335 revolutions per minute. Note: The term vertical implies a vertical main and gate shaft which extends above flood level to protect generator and electrics.

System 2

22 inches of head — 24 inch CMC-Fitz vertical Francis develops three amps at 130 volts or 72 amp-hours per day or 9.36 kilowatt-hours per day. Discharge is 520 cubic feet per minute at 70 revolutions per minute.

System 3

Three feet of head — 30 inch Trump vertical Francis turbine develops 35 amps at 130 volts or 840 amp-hours per day or 109.2 kilowatt-hours per day. Discharge is 2,358 cubic feet per minute at 96 revolutions per minute.

System 4

15 feet of head — eight-inch MacLeod-built CMC vertical Francis turbine develops 12 amps at 130 volts or 288 amp-hours per day or 37.4 kilowatt-hours per day. Discharge is 130 cubic feet per minute at 580 revolutions per minute.

System 5

Four feet of head — 27 inch S. Morgan Smith vertical Francis turbine develops 28 amps at 250 volts or 672 amp-hours per day or 168 kilowatt-hours per day. Discharge is 2,190 cubic feet per minute at 123 revolutions per minute.

System 6

10 feet of head — 12 inch CMC vertical Francis turbine develops 15 amps at 130 volts or 360 amp-hours per day or 46.8 kilowatt-hours per day. Discharge is 244 cubic feet per minute at 320 revolutions per minute.

Low Head Hydro Information

Getting info on low head hydro isn't easy. Virtually nothing of any technical merit has been published since 1940. Watch out for crazies and experts who try to reinvent the wheel. It is unnecessary and wrong-minded. It has all been done and done well. Go find the data. Rodney Hunt Manufacturing published some of the best information between 1920 and 1950. They also built great machines. They no longer build turbines. Their books are out of print. Find them in engineering school libraries or museums that specialize in early industrial technology. Turbine makers catalogs from 1880 to 1920 were in fact engineering manuals, some better than others. Look for them. I haunt the old bookstores. Go for it.[1]

Some Words of Encouragement…

Well people, I hope I've opened the door to stand-alone, low head hydro for a few of you.

If you really want the details you've got some long hours of research ahead of you. If you are determined to get on line, I wish you the best. Watch out, it is harder than building a house from scratch. It can be a real relationship buster. I believe it has as much merit as any effort at self-reliance one can undertake. Good Luck![2]

Editor's note: This article first appeared in *Home Power* Issue #23 (June/July 1991) and is reprinted with permission of the author.

Rolling Thunder

RICHARD PEREZ

When Stuart Higgs visited Hoover Dam at age nine, he dreamed he would someday make his own electricity from flowing water. Now 50 years later, Stuart and his family operate the biggest home power system I have ever seen. Two families, both with all-electric homes, are supplied by Stuart's hydroelectric turbine. With a daily output of up to 720 kilowatt-hours, Stuart's hydro could power ten average North American households, or over 50 energy-efficient households. And it cost about the same as an automobile, plus years of study, research and just plain hard work by Stuart.

Hydromania in Our Back Yard

Late one evening, Bob-O called to tell us that the winner of an international hydro competition lived not 30 miles from us. The Yreka, California, newspaper carried a story about a local man, Stuart Higgs, who had just placed first in an international competition to design and build the most effective hydro turbine runner. This competition, at the International Water Power Conference '91 in Colorado this summer, featured entrants from many nations and all large hydro players. A man in our back-yard skunked all the big-time operators and

took home first place with his $12 homemade hydro runner.

As you can imagine, we were very interested in meeting Stuart. Since the newspaper didn't give any access data, we tried *Home Power*'s subscriber database. Sure enough, Stuart was a subscriber. Armed with his address, we quickly got his phone number from information. We called and set up an interview. Here's what we found out.

The Higgs' Homestead

To the north of Yreka, California, the Shasta River flows from the 14,000 foot bulk of Mt. Shasta into the Klamath River and then into the Pacific Ocean. Along the river's way to the ocean, Stuart borrows some of its water for about a quarter mile and then returns it. Stuart's site is about seven miles from downtown Yreka and three miles from the nearest commercial electric lines.

Stuart has been a hydromaniac since his visit to Hoover Dam. He chose the site of his present home with hydro power in mind. Years of work finally became a hydro system on Christmas Eve 1989. Stuart's wife returned home to find their homestead brilliantly lit from top to bottom with Christmas lights. Stuart had switched the hydro on for the first time, and everything worked!

All-Electric Homes

I am not going to dwell on the specifics of the appliances powered by Stuart's system. This data is meaningless, and the list of appliances would fill pages. When a renewable source produces as much power as Stuart's hydro, there is no point in counting kilowatts. Stuart powers up two all-electric homes. Everything is run on electricity. Everything. Included are appliances we do not normally associate with renewable energy systems — big-time electric power slurpers such as electric clothes dryers (two of them at 5.5 kilowatts each), electric space heating via many baseboard heaters, electric hot water heaters, air conditioning, electric cookstoves, multiple refrigerators and freezers, dishwasher, trash compactor and myriad high-powered shop tools (like a 3 horsepower air compressor). All this and more are powered by Stuart's hydro. I noticed a wood heater in the living room and asked Stuart about it. He said they installed it as a backup heat source and have never used it.

In terms of electric appliances, the Higgs Homestead has just about everything you could imagine. When you own the power company, why not?

Stuart's Hydro Site

Stuart uses 1,200 feet of ditch to deliver water to his turbine. The head (or vertical distance that the water falls) in the system is 17 feet. The turbine cycles between 10 and 30 cubic feet of water per second (between 5,000 and 15,000 gallons per minute), depending on the water level in the river. On the dry August day we visited, the turbine was cycling about 12 cubic feet per second (5,400 gallons per minute) and was producing about 12 kilowatts of power.

Stuart made sure of the water rights on his homestead before he moved. His homestead holds water rights for 50 cubic feet per second. He tore down the old wooden flume that delivered water to the site, and replaced it with a large ditch. This ditch required both blasting and heavy equipment to construct. Stuart did the work himself with his D6 Cat, a crane and a backhoe.

The Fish Screen

The ditch delivers the water to the hydro through a fish screen. This fish screen is a marvel of design and function. A large area (about 6 feet by 20 feet), fine mesh, stainless steel screen prevents fish from entering the hydro. The screen is continually wiped by long brushes to keep debris from clogging it. Everything is automated and powered by electricity (what else?). The Shasta River is sometimes full of migrating fish. Stuart's screen works so well that the California Department of Fish and Game often bring ranchers, and others using river water, to see it. Whoever claims that small scale hydro turbines are a threat to fish hasn't seen Stuart's fish screen. The fish screen feeds the river into the turbine via a four-foot diameter pipe.

The Turbine

Stuart's turbine uses a horizontal axis, Francis-type reaction runner. The turbine was built by the Morgan Smith Company and rebuilt by Stuart. This unit is huge — about 6 feet in diameter, 15 feet long, and has a main shaft diameter of 4 inches. Stuart rates its output at about one kilowatt of power for each cubic foot of water per second fed into the turbine.

The turbine is belted up to a 30 kilowatt, 120/240 VAC alternator. This alternator makes 60 cycle AC power directly. Stuart's system uses no inverters or batteries, but makes its power as it spins, hence the name, Rolling Thunder. And thunder it does. The feeling of being in the powerhouse is indescribable. Up to a ton of water is roaring through the turbine each second. The deck of the powerhouse shudders under the force. There is no doubt to the senses that rolling thunder is harnessed within the turbine.

A Thomson and Howe hydro control uses five, six kilowatt shunt heating elements to keep the frequency of the alternator at 60 cycles per second. Stuart says that the frequency output of the controlled turbine is accurate enough to run standard electric clocks for months before they gain a few minutes. The Thomson and Howe control is capable of absorbing the full 30 kilowatt output of the turbine.

Stuart said that he is only using half of the turbine's runners because he is already generating more power than they can use. If the need should ever arise, Stuart could allow water to flow over the second runner in

A Primer on Hydro Runners

The business end of a hydro turbine is called a runner. The runner converts the moving energy of water into mechanical power by turning the output shaft. The runner is the interface between the world of flowing liquid energy and rotating mechanical energy. Hydro runners come in two basic types, those which operate in air and those that operate totally submerged in water.

Turbine runners that operate in air have the water sprayed onto the runner through the orifice. The stream of water moves through the air and hits the cups on the wheel. This impact turns the shaft. This type is called a Pelton or impulse turbine. This type of runner, one that operates in air, is most commonly used on microhydros like those made by Harris, Energy Systems and Design and Lil Otto Hydroworks.

A second class of hydro runners operate totally submerged in water. These turbines are like propellers, aircraft wings, helicopter rotors and the propellers on wind machines which operate by using an airfoil. The shape of the runner's (or airfoil's) blades is such that the surface area of one side of the runner is greater than the other side. The fluid motion across the runner creates unequal pressure on one side of the runner. This pressure is created because the water must move unequal distances across the unequal surface areas of the runner. The net result is a force, produced by water flowing by the runner, that turns the turbine's shaft. And all this happens totally submerged in water. This type of runner is called a *reaction* runner. Reaction runners are found in turbines manufactured by Canyon Industries, Almanor Machine Works and others. If you want more info, see a physics book under turbines.

the turbine. If he did this, then the system would produce about 50 kilowatts or over one megawatt-hour of power daily.

A Hydro Breeder

After touring the turbine, we visited Stuart's machine shop. Stuart uses hydro power to build, what else but, more hydros. Kind of like a breeder nuclear reactor without the glow-in-the-dark features. The turbines that Stuart makes are truly beautiful works of art. The reaction runner, shaft and other critical parts are constructed out of stainless steel and are finely finished. Stuart considers his home-built turbines to be his finest accomplishments, and is far prouder of them than his international first place award.

Hydro Doesn't Just Happen

You don't just wake up one morning and realize that you have big-time hydro potential. It's something that you plan and work a lifetime for. Just like Stuart did. Stuart's work has given his family energy self-sufficient homes that spare no convenience. And do no harm.

Stuart is a farmer. He has no formal training or experience in hydroelectric systems. He has no deeper pockets than most of us. His accomplishments spring from an intelligent and inquiring mind that isn't afraid of hard work. Stuart didn't have any hard figures about how much his system cost. He did the construction work and built or rebuilt most of the hardware himself. He did say that his hydro has produced power at less than one cent per kilowatt-hour since it went on line December 24, 1989.

When I spoke with Stuart, I saw the spark in his eyes that had become rolling thunder. He had nurtured a dream of freely flowing energy independence for 50 years and made it real. The world is his oyster. It's really hot today, so turn up the air conditioning, get some iced tea out of the reefer and find out what's on satellite TV. Nature is providing the power, and Rolling Thunder is footing the bill.[1]

Editor's note: This article first appeared in *Home Power* Issue #25 (October/November 1991) and is reprinted with permission of the author.

26

Handmade Hydro Homestead

BOB-O SCHULTZE

In 1975, Matt and Roseanne Olson's diesel generator died. Faced with an expensive repair, Matt figured this was as good a time as any to build the hydro plant which he'd been collecting parts for over the past four years. With the help of his good neighbor and friend Rod Ward, he set to work building a temporary fix for his generator problems. Eighteen years later, that "temporary" fix is still producing clean, renewable electricity. The old diesel plant has long since been traded for spare parts.

Location is Everything

Matt and Rosie live at the confluence of Methodist Creek and the Salmon River in Northern California. Their homesite lies in the western part of the Klamath National Forest. This river corridor is still very scenic and colorful, despite placer mining before and during the 1930s, the ravages of two major fires and extensive road building and timber cutting by the Forest Service over the past 30 years. The beauty of the Salmon River Country is due in no small part to reclamation by folks like the Olsons and small scale miners along the river

who have adopted an attitude of stewardship and living with the land rather than from it.

This country is a dandy place for low impact microhydro because of the many creeks and springs which feed the Salmon River. The nearest electric utility lines are 23 miles away through one of the steepest, most picturesque river canyons in California. Even if the locals wanted utility power — which they don't — and even if the utility was willing to provide it — which they aren't — the cost and visual impact to the area would be prohibitive. So it's a case of using what Mother Nature provides on-site and learning to live on her terms. Not a bad philosophy for all of us.

It's the Water

It takes a fair amount of water to run an overshot wheel. Matt and Rosie's hydro and irrigation water comes from a recycled mining ditch which flows along the back edge of their property. A weir — that is, an adjustable gate — at the intake from Methodist Creek determines the flow into the main ditch which feeds the ram pump, flood irrigation check ditches and, finally, the overshot wheel itself. This weir restriction serves two functions. First, it keeps out the anadromous fish which use Methodist Creek as a spawning ground. The existence of large redds, or spawning beds, at and above the intake attest to its low impact. Second, the weir limits the flow of water into the ditch during high water periods which would cause erosion of the ditch banks and overspeeding of the wheel.

All along the ditch, trees, bushes and a giant thicket of blackberry vines drink their share of the water. How much do they take? "About 500 watts per hour worth during a hot summer's afternoon," according to Matt who can watch the power drop as the temperature climbs. "The plants, grass and especially Rosie's flower garden come first where the water is concerned," said Matt. "In low water years we may have to shut the wheel down for up to three months because they get priority." After the water leaves the wheel, it flows down through an overgrown mining tailrace and into the Salmon River, none the worse for wear.

Recycle, Reuse, Rebuild

Nearly every part of Matt and Rosie's hydro plant has been reincarnated after dying as something else. The hub and main shaft of the 13 foot diameter overshot were part of a 24 inch Pelton wheel with the cups removed. The one-inch steel rods radiating from the hub came from a scrap metal pile. The floor of the buckets and the 18x12 inch deep buckets themselves were painstakingly cut from an old dump truck body and individually welded into place. Most of the pulleys, sprockets and jack shafts for the speed multiplier gear train were bartered or scrounged from deceased

mining and farming machinery. About the only things purchased new were the bearings for the main and jack shafts and the V belts. In 1975, those items cost approximately US$250. To replace them all at today's prices would run about US$550.

The 1,800 revolutions per minute, 2.6 kilowatt Kurz-Root AC generator came out of a 1940s military portable GenSet. Matt replaces the brushes about twice yearly and trues the commutator and slip rings every couple of years. He tells a great story about filling a missing commutator insulator with JB Weld™, which turns out to be nonconductive, and turning it down on a lathe. "That was over a year ago, and the dang thing still runs great."

The Hydro Plant

The total head of the Olsons' hydro system is about 14 feet. This includes the pitch on the wooden chute which acts as a nozzle and the 13 foot diameter wheel itself. At full output, the Olsons' hydro system produces 2,500 watts (2.5 kilowatts) of 120 volt, 60 hertz AC power. This hydro produces a whopping 40 kilowatt-hours daily.

The wheel uses one cubic foot per second of water and turns at 12 revolutions per minute while under load. Through a system of jack shafts that would do Rube Goldberg proud, the speed is increased to approximately 2,000 revolutions per minute. Maintenance consists of greasing all the bearings weekly and knocking some of the ice buildup off the wheel during the coldest part of the winter.

Living with an AC Hydro Plant

There are no batteries, inverter or controls that we normally associate with a stand-alone renewable energy system. It's rolling thunder, and you have to use what you produce — one way or another. Increasing the wattage loading past the generator's output will cause low voltage brownouts and lower than 60 hertz power frequency. Decreasing the load on the generator will cause the wheel to spin faster, subjecting all the appliances and lights to a high voltage and frequency condition.

Matt and Rosie use the simplest form of manual load management. They leave the lights on. Rosie jokingly says, "I've had friends who don't know about the system come to the house and tell us 'That must have been a hell of a party last night for you to go to bed and leave all the lights on!'" They also keep an eagle eye on the voltage and frequency meters mounted in the kitchen between the sink and the refrigerator. "We just always glance at it as we go by,

it's not even something we think about anymore," according to Rosie. What better place for metering than the highest traffic area in the house? It's a great place to install the system instrumentation in any RE-powered home.

This kind of load management would be very difficult — even dangerous — with a high revolution impulse wheel system like a Pelton, but the overshot wheel turns at only 12 revolutions per minute. Consequently, it takes a while to change the revolutions of the generator, and hence change the voltage and frequency, one way or the other: time enough to turn an appliance on and some lights off without a mad dash for the switch.

Hydro-Power Appliances

Matt and Rosie's system powers all the electrical appliances they need. Like most folks

---

The Physics of Falling Water

Gravity powers the overshot water wheel. The water falls, and as it falls it does what physicists call work because its mass is accelerated by the pull of gravity. The overshot water wheel converts the energy of the water's falling mass into mechanical power at the wheel's axle. The amount of power available for conversion depends on two factors: the amount of water per second flowing over the wheel and the vertical distance that the water falls. Here's the equation:

P equals W times H divided by 0.7376

where P equals the available power in watts. W equals the weight (in pounds) of water flowing over the wheel per second (a gallon of water

weighs about eight pounds; a cubic foot of water weighs about 64 pounds). H equals the distance (in feet) that the water falls.

Most overshot wheels only capture the water for 120° of their rotation. With 120° rotation, H is equal to 1.5 times the wheel's radius. The 0.7376 is a fudge factor to make the power unit come out as watts rather than foot-pounds per second. From the equation, two facts are obvious to every hydromaniac: the more water flowing over the wheel per second the better, and the longer distance that the water falls the better. These two factors, flow and head, determine the power potential of all hydros regardless of type.

powered by hydro, they were vague about their power consumption. When your concern is keeping the hydro's constant power output under control, things like lights burning all night are common. Matt and Rosie power lighting, a satellite TV system with color TV, Matt's machine shop full of power tools and a slew of kitchen appliances. Cooking and water heating is fueled by propane.

Using Water to Get Water

Rather than using electricity to run a pump for the house water, Matt and Rosie use a 40-year-old Rife™ ram pump. The Rife is fed through a two-inch diameter pipeline dropping about 20 feet from the ditch into another ancient mining tailrace. They've been using the Rife continuously for the past 24 years!

The ram pumps against a large pressure tank which also feeds a couple of sprinklers for the lawn during the summer and an open overflow line in winter. Matt figures that the Rife produces about 10–15 gallons per minute. By keeping track of the amount of water being used continually, they can maintain about 25–30 pounds per square inch of pressure in the tank…plenty for most household uses.

Matt has modified the captive air tank on the ram pump by adding a couple of small petcocks. These valves make the weekly chore of draining the water and reestablishing the *air cushion* in the pump just a five minute job. The only other maintenance Matt has performed during the ram pump's 24 year tenure is replacing the rubber seals and gaskets "every five years or so." After buying the first set of replacement gaskets, Matt has been making his out of a section of discarded rubber conveyor belt.

Conclusion

It wasn't all that long ago when the Olson's lifestyle and philosophy of recycling, rebuilding and reusing was considered pretty backward. Today, most of us have caught on to the *Three Rs* in one way or another, and it turns out the Olsons are pretty forward after all.[1]

Editor's note: This article first appeared in *Home Power* Issue #37 (October/November 1993) and is reprinted with permission of the author.

Choosing Microhydro—
Clean Electricity in the Outback

JEFFE ARONSON

It was a dream of mine since my teens — building my own microhydro-powered homestead in the mountains, on *my own* river. And finally, here it was. We had the money, we found the right property and all was go. When you've spent close to A$25,000 to power your home, you learn some lessons. Here's some perspective for the next dreamers, to prepare for the good, the bad and the ugly. In spite of our problems, we have what we consider to be the cleanest, most cost-efficient, best source of electricity in our remote mountain setting. Being on the grid was impossible, since we are over six miles from the end of the line, with only a couple of dozen inhabitants in the vicinity. So the choices were solar-electric panels, hydro, diesel or petrol engine/generator, wind turbine or no electricity at all.

Water Is the Answer

We wanted the modern conveniences we were used to, but couldn't see the point in living next to a beautiful river constantly murmuring in the background but drowned out by the sound of an engine/generator. The environmental consequences didn't appeal either, and I figured I had better things to do with my time than constantly working on a greasy, oily, cantankerous engine.

When we found our valley, we visited some soon-to-be neighbors who had a combination solar/hydro system and who recommended their installer to us. They had a hundred-year-old water diversion and used that water for their high head hydro. They had enough electricity for basic living, but didn't seem to have much left over for laundry, power tools, TV etc. We contacted the installer, since he'd put together four local hydro systems, and everybody knew him and was happy with his work.

Our first real lesson in hydro, though I'd read quite a bit about it in alternative living books, was the difference between high head and low head systems. *Head* is the height of fall of the water. To run a hydro system and calculate the potential output, you have to know the head and the flow, including the lowest and highest flows expected during the year. You can get sufficient power from a combination of high head and low flow — say, three to five gallons per second dropping from a spring 65 feet or more above your hydro unit. Or you can get similar output from a low head, high flow system like ours — a constant supply of between 30 to 50 gallons per second dropping only 6.5 feet.

Our installer came out to check our little waterfalls out and pronounced the site doable. He'd been in the area before and had two other units in the valley (high head ones). He'd never before seen a site he'd guarantee in the main river, since the danger of floods ripping out the turbine was too high. Our site, however, had rocky banks and eddies that seemed as if they'd protect the works from major floods. He confidently told us to expect "at least" a 400 watt output.

Load Analysis

For those first contemplating homemade electricity, there are some things to know. To figure how much energy you'll need in your household, you have to put together a chart of all the appliances you want to use, their rate of energy consumption and how long you'll use them each day. Like a budget, you should be pessimistic and overestimate how long and how often you'll use things, just in case.

Using these calculations, you can then figure out how much energy your household will need on an average day and on a high-use day. All households are different, and if you're moving from being on the grid to homemade electricity for the first time, you'll find the learning curve steep in the beginning. Over time, you'll see your roles as *power plant managers* evolve — your family will have to be involved as well! Things get easier and more like habits later, but at first you might be a tad overwhelmed.

Whatever system you use, from solar-electric to gas generator to hydro, they all take time, thought and effort. For some, a very

Aronson Loads	
Item	**Average watt-hours per day**
Assorted always-on and phantom loads including computer, answering machine, stereo, TV, composting toilet fan and clocks	1,382.40
Ceiling fan (summer)	1,080.00
Larger power tools	720.00
Washing machine	211.20
Microwave oven	192.00
Freezer	190.08
Small power tools	180.00
Satellite TV and VCR	151.20
Radio, stereo, CD player	144.00
Refrigerator	144.00
Kitchen lights, fluorescent	86.40
Hot tea water jug	86.40
Living room lights, fluorescent	28.80
Mixer, juicer, blender, etc.	26.88
Blow dryer	23.76
Toaster	18.00
Bedroom lights	14.40
Outdoor lights	13.44
Bathroom lights	7.20
Vent fan	4.9
Computer printer	0.38
Total average watt-hours per day	**4,705.34**
(kilowatt-hours per month)	**(141.15)**

Jeffe Aronson

simple system will do since the requirements will be small. We wanted a washing machine, stereo, reasonably sized TV, computer, microwave, water jug, toaster, fridge and freezer, vacuum and to use power tools, all without the hassle and noise of a generator. So we needed lots of electricity, which translates into lots of money and sweat! Having done our load chart, we saw that a 400 watt output, or more than nine kilowatt-hours a day, was perfect. Remember, hydro is 24 hours a day, as opposed to PVs, which obviously only produce when the sun is shining. We were ready to get started.

Permits

Before we could start on the actual work, we had to deal with the permits and also the need to alter our beautiful little river. In terms of permits, "everybody" suggested we just do it and let the bureaucrats find it if they could. Despite my natural tendency to hate the bastards (just like everyone else), I had worked for years with politicians and bureaucrats and knew two things about them (besides what we all experience and know, of course…). First, if you work with them and don't show anger or frustration at some of their quirks, establish a good working relationship, show them respect (whether you really think they deserve it or not), "butter them up," so to speak, you just might get what you want out of them, sooner or later. Second and perhaps more important, since they do have the law on their side, if they catch you in a lie or doing something against the rules or trying to pull the wool over their eyes, you are screwed. Royally.

I didn't want to deal with that weight on my shoulders, ruining my feelings about my home, so we decided to go through the permit process. Indeed, after a little while and a

---

Calculating Hydro Power

Have a potential hydro site and want to estimate the power available? If you know your head and flow, you can do a very simple calculation for a rough estimate of the continuous output in watts. Just multiply the net head in feet by the flow in gallons per minute, and divide by an adjustment factor. Use a factor of nine for an overall efficiency of 59%, typical in AC systems. Use a factor of 10 to 13 for an overall efficiency of 53 to 41%, typical in DC systems. If you don't know your net head (which includes friction losses in the pipe), use the total head and then take 10 to 20% off the total.

Examples:
- 7 feet of head times 6,000 gallons per minute divided by 9 equals 4,667 watts
- 70 feet of head times 600 gallons per minute divided by 9 equals 4,667 watts
- 700 feet of head time 60 gallons per minute divided by 9 equals 4,667 watts

- 5 feet of head times 1,000 gallons per minute divided by 10 equals 500 watts
- 50 feet of head times 100 gallons per minute divided by 10 equals 500 watts
- 500 feet of head times 10 gallons per minute divided by 10 equals 500 watts
- 2 feet of head times 700 gallons per minute divided by 13 equal 108 watts
- 20 feet of head times 70 gallons per minute divided by 13 equals 108 watts
- 200 feet of head times 7 gallons per minute divided by 13 equals 108 watts

Remember that all the energy you get from a batteryless AC hydro system has to be used at the same time it is produced. In battery systems, you can store unused energy for use later when you actually need it.

modicum of paperwork and approvals by all sorts of agencies from environmental to Aboriginal groups to parks and onwards, our hydro system was approved.

This was fortunate, since only a couple of years later a visiting fisherman, who had fished this section of river for decades and considered it his own, came upon our works, was outraged and rather than discuss it with us, sent in a complaint to the water catchment authority. They sent the representative with whom we'd dealt originally. He thankfully found that we'd done what we'd said we'd do and even felt the works to be "very discreet." He was visibly shocked to learn that we planned to cover what we could with wood to make it fit in even better (since done). We now have a letter of approval to this effect, just in case...

Loving and Using the River

As for the river alterations, this actually caused us the most anguish. I've been a river guide all over the world for nearly 30 years. I love rivers. They've nourished me and healed me, not to mention providing my living and years of fun and excitement my whole adult life. Contemplating moving things around and then pouring concrete was agonizing. Twice we begged the installer to consider doing the project with pipes rather than the water channel. He refused.

We considered nixing the whole project, but when we compared it with the environmental damage, visual and noise pollution of a generator, we balked. When we considered the cost, even after rebate, of enough solar-electric panels for our lifestyle, it was prohibitive—

much more than the A\$15,000 to A\$18,000 estimate for our hydro.

We ended up justifying things to ourselves by committing to making the least possible change to the river's edge, not affecting the course of the river in any way and making sure that it remained fishable and kayakable (even though it is only high enough to kayak for maybe three to seven days a year, at best). We also planned to make the works as inconspicuous as possible, to the extent our ingenuity and pocketbook allowed.

Utility electricity, which we'd been on all our lives, results in air pollution, hundreds of miles of rivers drowned under huge reservoirs, hundreds of square miles of open coal pits and mines and more. Our commitment to minimal impact seemed like a better bet to us. We've since had a few neighbors, as well as some river running colleagues, admonish us for what we did. Expect this, especially if you're building a hydro in a formerly pristine area. But those who have had the courage to discuss our options with us agreed in the end that, considering all the alternatives, we were choosing the lowest-impact option. These on-grid folks had to admit that we had affected the environment far, far less than they do just by having a hot cup of tea or a cold beer.

Custom-Built Turbine and Civil Works

Our low head hydro unit was going to be built specifically for our site by Peter Barrett's Platypus Power in Bright, Victoria, Australia. Our first challenge was to create a water channel for the intake of water into the turbine. High head units only require an intake screen and a pipe—pretty simple.

Technical Specifications

System type: Off-grid microhydro
Location: Victoria, Australia
Hydro resource: 6.5 Feet head, 30–50 gallons per second flow
Production: 180 AC kilowatt-hours per month average

Hydro Equipment

Turbine: Custom
Runner: Custom by Platypus Power, six-inch hub with ten, two-inch aluminum blades
Alternator: Fisher and Paykel Smart Drive, 24 VAC nominal, 750 watts maximum
Measured performance: 13.5 amperes maximum at 500 revolutions per minute, 3 gallons per second
Hydro overcurrent protection: 30 ampere breaker

Balance of System

Charge control and dump load: AERL Maximizer
Inverter: Solar Energy Australia 2500, 24 VDC nominal input, 220 VAC, 50 hertz sine wave output
System performance metering: Plasmatronics PL-20
Engine/Generator make/model: Honda 6X200, 6.5 horsepower

Energy Storage

Batteries: 20 SAFT, NiCad, 440 ampere-hours, 1.2 volt cells battery pack: 24 VDC nominal, 440 amp-hours total
Battery/inverter disconnect: 250 amperes

Jeffe Aronson

The turbine is over six feet tall and ten inches in diameter. A shaft connects the alternator on top to the runner below. Below it is a Glockmann water-powered pump.

However, our installer wouldn't guarantee our hydro with a piped inlet, since he'd seen this river in flood and was afraid the pipe would get ripped away downstream, tearing out other bits along with it. He insisted on a water channel intake from one pool in the middle of our little waterfall to the pool below the last drop — a total fall, or head, of just under 6.5 feet.

This required a great deal of moving of rocks along a 50 foot section of riverbank — by hand. It was probably over five tons, by my back's estimate. After this work, we had to pour a concrete retaining wall or weir, where the water would pool at the end of the new concrete channel or *water race*. It would then drop into the turbine through the intake pipe built into the bottom of the weir.

This concrete weir is about 24 square feet per side and poured to fit snugly in the natural

Aronson Hydro System

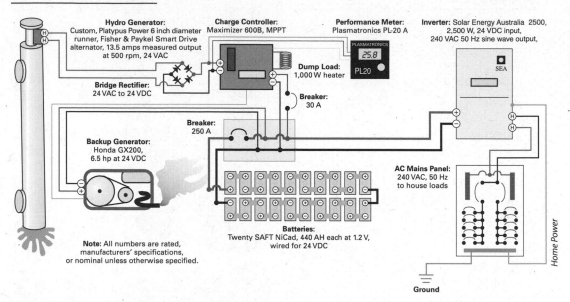

Hydro Generator: Custom, Platypus Power 6 inch diameter runner, Fisher & Paykel Smart Drive alternator, 13.5 amps measured output at 500 rpm, 24 VAC

Bridge Rectifier: 24 VAC to 24 VDC

Backup Generator: Honda GX200, 6.5 hp at 24 VDC

Charge Controller: Maximizer 600B, MPPT

Performance Meter: Plasmatronics PL-20 A

PLASMATRONICS
25.8
PL20

Dump Load: 1,000 W heater

Breaker: 30 A

Breaker: 250 A

Breaker: 250 A

SEA

Inverter: Solar Energy Australia 2500, 2,500 W, 24 VDC input, 240 VAC 50 Hz sine wave output,

AC Mains Panel: 240 VAC, 50 Hz to house loads

Batteries: Twenty SAFT NiCad, 440 AH each at 1.2 V, wired for 24 VDC

Note: All numbers are rated, manufacturers' specifications, or nominal unless otherwise specified.

Ground

Home Power

gully between two very large protective rocks. The concrete water race is a 49 foot long diversion, built into the rocks along the bank, about 1.6 feet deep and 8 inches wide, channeling water along one bank of the river to the weir. The space between the water race wall and the bank averages about 1.6 feet. The water then pools behind the weir. The 10-inch intake pipe in the bottom of this weir takes the water into the turbine itself.

All of this was done on some rather uneven and rocky terrain, at the bottom of a very steep and rocky 26 foot bank, accessible only as far as its top by four-wheel drive. Needless to say, we had our work cut out for us. In fact, I soon learned to describe this project as the hardest job I'd ever done.

Pouring Concrete

After digging the channel in the bank, we hired a contractor to build the forms for pour-

ing the concrete. I'd worked with concrete before, but forming over very rocky and uneven terrain, in a very remote spot, where it needed to withstand tons of water flooding over a waterfall for several weeks a year convinced me to seek someone with real experience. Our installer made it clear that he wasn't a concrete man and wasn't really interested in doing the civil works, preferring the electrical end of things for his stimulation. He told us how long, how wide, how high and how level he wanted the channel and left us to our devices. "Let me know when you're ready for the hydro."

Thankfully, our neighbors and the son and daughter of the local concrete contractor helped us with the pour. It involved tons of sand and gravel, bags of cement, a petrol-driven motorized concrete mixer, a fire pump with intake hose in the river for water, a corrugated iron chute, shovels and buckets. It was a lot of sweat for a half day, after getting the equipment to the end of that very steep four-wheel-drive track.

Turbine Installation

After our preparation was done, the turbine was delivered, and two neighbors and I dragged it down the hill and put it into place. Basically, it's a 6.5 foot long by 10 inch steel pipe. The top end is capped except for a hole drilled for the shaft and shaft housing. A ¾ inch stainless steel shaft goes through the pipe attached to a housing for the generator at the top. Near the bottom, the shaft goes through a fitting that the vanes and propeller blades (made from air fan hubs and blades) are attached to.

The vanes aim the water onto the angled propeller blades which turn the shaft. A small, capped flange for blade access is on one side of the pipe near the blades at the bottom, with another flange to attach the intake pipe to near the top. A butterfly valve between the intake pipe and turbine allows us to turn the water, and thus the power, on and off. An air bleed valve was attached to the top cap.

Fortunately, everything fit perfectly with the intake pipe, which was already cemented in place in the weir. Now it was the installer's turn, while I continued building our house, complete with water tank and plumbing, woodstove and access roads.

The installer came out, installed the inverter and controller and hooked up the battery backup system. He then installed the generator (a rewired Fisher and Paykel washing machine motor) on top of the turbine, and we fired the thing up for the first time. After all that work, fear and agony, we were so happy to turn on our first light. It worked!

Problems

As it turned out, the highest output the installer could coax out of the turbine was about 324 watts, or not quite eight kilowatt-hours a day. We tweaked it for months, but that was it. Mostly, it averaged between 250 and 290 watts, but since we had electricity (and we didn't have much choice anyway), we took a wait-and-see attitude.

We had a number of other problems. Leaf litter and stick debris constantly clogged the blades. To avoid an expensive, time-consuming and visually destructive retrofit, we installed three separate stainless screens, with ½ inch holes, along the water race. We now take a daily walk down to the river to clear them of debris, but it eliminates the need to freeze or nearly drown while clearing the blades.

We also changed the blade angle to allow more debris to pass through. The blade access hole was built too small for me to fit into, but my wife Carrie was worried about more retrofits and agreed to do the blade cleaning. The water flow through the turbine was very turbulent due to the lack of a cone above the vanes to smooth and direct the flow, and another below the blades to reduce cavitation as well as the lack of a vacuum outlet. Fixing this will require a major rebuild. Air pockets at the top of the turbine were fixed with an air bleed valve.

Our controller didn't give us some information regarding amp-hours and daily histories, which we wanted for system evaluation. We got a new one installed. The steel intake pipe and turbine are rusting faster than we'd like, and weren't coated with epoxy.

On our next rebuild, we'll coat it for longevity. A backup engine/generator wasn't installed, so when we had floods or did maintenance, we had no reserve past the minimal battery bank. We've since installed a backup generator to top up the batteries when the turbine is off-line.

Mixed Success

After three years experience, and much studying and discussing with experts, we've had little better than mixed success with our power plant. It does indeed work, but after much analysis by several microhydro experts, it should be producing two or three times as much power with the head and flow we have. It should also be far easier to maintain. We've even asked our installer to fix the bugs in it, offering to pay for the labor and material, but he's simply not interested. Living in Australia, there isn't much opportunity to find others to help us. After three years of research, phone calls, e-mail and photographs, we find that the few individual companies that work with hydros either don't do low head systems or live too far away.

We're glad we've gone hydro for many reasons. If and when we gather the money and fortitude and find someone to assist us, we believe we could improve our system to provide more output, even during floods, and eliminate most of the hassles as well. In the meantime, we love our home and the silent, clean, near-constant smooth electricity our system gives us.

We love saving literally thousands a year on ever more expensive gas (propane in the US) by having an electric freezer and fridge. We love simply turning on the circular saw, hair dryer or toaster without hearing the generator out back. It does require a quick glance at the battery status and power generation. Several of these items require more than we can produce at once, necessitating the use of stored energy from the battery bank, after which the hydro replaces it over a short period of time.

Listen to the River

Realizing your dreams sometimes takes you down paths you didn't expect, and often requires a lot more work than you'd planned. We've made some mistakes along the way, some of which are fixable, some not. But we're proud and thrilled that we've persevered and overcome. In the end, we have come darn close to achieving our goal of environmentally friendly, efficient, reliable homegrown electricity.

If you're on the same path we are on, consider how much money you have to spend, how many appliances you feel you *must have*, how much time and energy and strength you want to put into energy management and take it from there. Whatever your choices, make sure you take time out to listen to the river.[1]

Editor's note: This article first appeared in *Home Power* Issue #101 (June/July 2004) and is reprinted with permission of the author.

Advice Column

Virtually all types of hydroelectric systems have one major problem in common: debris. You should definitely predesign an efficient and perhaps automatic debris deflector/eliminator. Do not be talked out of this under any circumstances. Some very ingenious ways — some expensive, some cheap as dirt — have been developed over the years to do this job. But once you've built the inlet pipes or canal, it may very well be too late.

You have two choices: Get a firm prediction of the expected output in writing or expect less than predicted. Be very suspicious of potential installers unwilling to back up their estimate by writing it down. And when you get an estimate in price from an installer, add at least a third to the price to make it more realistic.

Expect to spend more time and energy than you would on a solar electric installation that has no moving parts and less of Mother Nature to deal with. Our hydro system has been about the same amount of work over time as a generator, though less greasy, dirty and noisy.

Plan for the worst case scenarios. In terms of energy needs, get the next bigger size inverter to the minimum you need. Otherwise, if you're having a party, or guests or a big project, your inverter may kick out or even fry and leave you back in the stone age for a time.

Steel rusts when in constant contact with water. Unfortunately, stainless steel is expensive and can be brittle. Numerous coatings are available for the innards of your turbine and intake. Coat your exposed steel before you install or expect to have to replace it sooner rather than later.

If the thought of being without electricity for even a week scares you, include a backup generator to charge your batteries right from the start. This eliminates the need to endanger your life to fix a problem or clear out debris during a flood or electrical storm. Just fire up the generator and wait it out. It is also a stress reducer during maintenance.

Check out microhydropower.net, a website for microhydro folks from all over the world. Join the discussion forum. They accept requests for information, guide you to professional websites, offer advice and have a great archive with drawings and photos of things like debris-free intake systems and installations. These folks pored over my photos and descriptions, asked questions and helped me to not only understand my system but to figure out how to improve it when I'm ready. They are people who have built numerous systems in the US, India, Europe and elsewhere. Their help was invaluable.

28

Water Rites—
A Microhydro Evolution

JEFFE ARONSON

Every morning, before brushing teeth or having a look at the weather outside our Victoria, Australia, off-grid home, my wife Carrie or I pad downstairs to the battery room to check the meters that monitor our electrical system. The reading determines whether we use the propane stove or the electric jug to heat water for our tea. Since our microhydro plant upgrade, it's usually the jug.

This ritual has become a part of our daily lives, like making the bed: a quick look at battery voltage and power from our microhydro system. It used to be a stomach-churning moment for us. Our previous microhydro turbine was dysfunctional too often due to its poor design, which meant a trip down to the river to clean the turbine blades—an uncomfortable and sometimes life-endangering task. Since replacing that experiment with an Energy Systems and Design (ES&D) LH1000 low head turbine, complete with a prototype leaf-mulcher and new PV modules to back up the hydro, off-grid life's simply blissful.

Off-Grid with Comfort
Living in the most remote part of the Victorian Alps, 9.3 miles from the electric grid, has not

reduced my appreciation for flipping a switch rather than filling and lighting kerosene lanterns, using circular saws instead of hand saws or for using other time-savers like toasters and microwave ovens.

Microhydro turbine manufacturer Paul Cunningham and the author work on attaching the penstock to the stand box.

Jeffe Aronson

195

Hydro System Setup

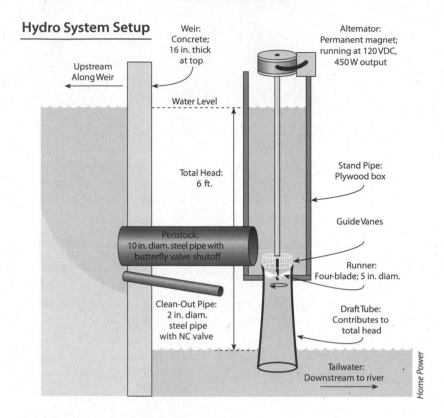

Weir:
Concrete;
16 in. thick
at top

Alternator:
Permanent magnet;
running at 120 VDC,
450 W output

Upstream
Along Weir

Water Level

Total Head:
6 ft.

Stand Pipe:
Plywood box

Guide Vanes

Penstock:
10 in. diam. steel pipe with
butterfly valve shutoff

Runner:
Four-blade; 5 in. diam.

Clean-Out Pipe:
2 in. diam.
steel pipe
with NC valve

Draft Tube:
Contributes to
total head

Tailwater:
Downstream to river

Home Power

Up and running — the first power output test

Jeffe Aronson

These days, although I'm an active 53 years old, I'm not content to spend my time chopping wood for cooking. Nor do I care to write by lantern light or pull a coolish beer out of the river. Instead, I want plenty of free time to go trekking, kayaking and skiing. Plus, I like my beer ice cold!

For the first few years, we had big problems with our original, locally engineered hydro plant (see Chapter 27). We feared that we'd made a huge mistake, as our time, money and energy was sucked into the hydro like leaves. Friends and neighbors shook their heads at our folly. But our recent turbine upgrade has markedly reduced our power plant management needs, freeing several daily hours for ourselves, and now our off-grid life is good.

Microhydro Madness

It wasn't until after we installed our first contraption that we discovered it was a manufacturer's experiment in low-head axial turbines. Its many 90° angles impeded flow, making it inefficient. Its access port to the blades was too small for an average-sized adult, and usually under water, making debris clearing a freezing-cold, often dangerous nightmare. We kept a cardboard *turbine box* near the back door stocked with a dry suit, goggles, snorkel, life jacket and safety rope. No kidding.

The turbine's poorly designed blades often came loose and rotated to the wrong angle, either jamming the unit or causing the turbine to lose a lot of power. The complete turbine weighed in excess of 441 pounds, making its removal a time-consuming process that required a winch.

Another unfortunate problem was that we weren't informed to build the intake with debris screening in mind. All hydro plants have to contend with leaves, sticks and other detritus. Because of the turbine and intake's design, we had to fiddle and mess with the turbine often, sometimes as much as three times a day. In contrast, a well-designed turbine intake should require a screen- or blade-cleaning perhaps once a week or, during summer low flows, once every few weeks. Unfortunately, when we first built the weir, we did not understand the importance of a debris-screening strategy. Proper design greatly minimizes, or can even eliminate, manual debris removal. Including screening in the original design would have

been easy, but now such an addition would be more difficult.

Our original turbine's 350 watt maximum output usually lasted only an hour or so after debris was cleared. After that, frequent checks of our Plasmatronics PL20 display would show the power declining steadily until we'd take our next forced march down the steep hill to reclean the turbine blades and intake screens. I dreaded the next flood or power drop, and dreamed of just going back to loving whatever mood the river was in. That cranky turbine cost us A$6,000 — three times more than the new ES&D LH1000 we replaced it with. Our naiveté and rush to build the hydro system cost us dearly — in money, time and energy. On the plus side, we learned heaps about microhydro systems and renewable energy and had in place all the other balance of system equipment — wiring, batteries, inverter and regulator — for a new turbine. But we'd finally had enough: after five years of struggle, our patience and nerves were at an end, so we decided to replace the flawed turbine.

Through a microhydro e-mail list-server group (see endnotes), we received advice from several folks. Paul Cunningham of ES&D in Canada read my *Home Power* article from 2004 and felt that one of his turbines would work for us. The LH1000 turbine he recommended was a fraction of the size of our original turbine, weighing about 44 pounds. It has a strong, cast bronze runner/propeller and could be made to offer blissfully easy access for unclogging.

We decided to give it a try and ordered one. Though Paul offered excellent support and advice over the phone and Internet, we invited him to visit us in Australia and help with the installation, and he accepted.

New Nuts and Bolts

With water flows of at least 2,000 gallons per minute, even during extended drought, and an average six-foot head, the new turbine was projected to provide at least 450 watts. Compared to the old machine, the new turbine required about 800 gallons per minute, so there was enough water to install two of the ES&D machines if need be.

It took half a day to remove the old turbine, and six people and a truck winch to drag the behemoth up the hill where it remains lying in the long grass like a carcass. I've since sold its generator — a washing machine motor — for a couple hundred bucks, assuaging my need to recoup at least something from the beast.

It took just another day and a half to nut things out at the river with Paul. We built the plywood box that channels the water to the turbine and carried it down to the river, where we bolted it to the intake pipe. After making some adjustments, we slipped in the draft tube, clamped the new turbine to it and opened the butterfly valve at the inlet. Just like that, we were up and running, and everything worked great on the first try! What a relief!

Sticks and Stones

We let our new turbine run for the first summer as-is, experimenting and observing. It worked well, regularly putting out about 450 watts. The debris problem remained, but was greatly diminished. With the old turbine, sticks, stones and leaves worked their way to the blades. These would either clog the blades, or a stone or stick might even break one. Removing the clog meant the water torture of shutting down the turbine, opening the intake (which was often below water level), reaching and contorting to get an arm down to the blades and grabbing out the glop bit by bit — all the while kneeling chest-deep in freezing water. Not fun.

But the ES&D turbine needs only minor cleaning. The new and improved muck-out process simply involved shutting the intake butterfly valve, removing one wall of the box and sticking a finger or two through the vanes to wipe the tiny leaves and algae off the blades — all at eye level, high and dry above the water line. The shape of the inlet vanes prevents sticks from reaching the blades. Although stones can still get drawn into the box, they drop safely to the bottom, awaiting removal the next time the access port is opened. The concern became small leaves and algae, which drape themselves over the blades, slowing the turbine and reducing power.

Through the microhydro listserv, I communicated with Michael Lawley of Eco Innovation in New Zealand. He'd been having similar debris problems with his Vietnamese-made, low-head unit, which he'd solved by inventing a simple leaf-mulcher. Paul and I decided to adapt this great idea to our turbine.

The mulcher is a little piece of plastic with its end shaped to mirror the blade tops. It extends down below the inlet vanes about 0.04 inch. Centrifugal force keeps the leaves at the outside edge of the blades, so the mulcher does not need to stick out very far. As the four blades spin, each passes by the mulcher 1,500 times a minute, which knocks or slices off the buildup. We had the mulcher in place all last summer and never once had to clean the

Aronson Hydro and PV System

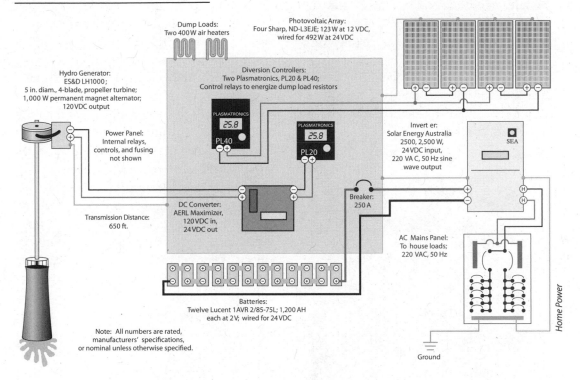

Dump Loads:
Two 400 W air heaters

Photovoltaic Array:
Four Sharp, ND-L3EJE; 123 W at 12 VDC,
wired for 492 W at 24 VDC

Hydro Generator:
ES&D LH1000;
5 in. diam., 4-blade, propeller turbine;
1,000 W permanent magnet alternator;
120 VDC output

Diversion Controllers:
Two Plasmatronics, PL20 & PL40;
Control relays to energize dump load resistors

PLASMATRONICS
25.8
PL40

PLASMATRONICS
25.8
PL20

Invert er:
Solar Energy Australia
2500, 2,500 W,
24 VDC input,
220 VA C, 50 Hz sine
wave output

SEA

Power Panel:
Internal relays,
controls, and fusing
not shown

DC Converter:
AERL Maximizer,
120 VDC in,
24 VDC out

Breaker:
250 A

AC Mains Panel:
To house loads;
220 VAC, 50 Hz

Transmission Distance:
650 ft.

Batteries:
Twelve Lucent 1AVR 2/85-75L; 1,200 AH
each at 2V; wired for 24 VDC

Note: All numbers are rated,
manufacturers' specifications,
or nominal unless otherwise specified.

Ground

Home Power

blades! Plus, instead of reducing the turbine's power, its output actually increased!

Now, our cue to clean the new turbine kicks into gear when the meter shows the turbine output dropping to between 10 and 11 amps (from 13 to 15 amps normally), which, in summer or winter, can occur after several weeks or, in autumn, after several days, instead of just a few hours. One of us strolls down to the falls and turns off the intake butterfly valve to drain the chamber. Then the valve is reopened a crack to direct a high-powered water jet onto one side of the inlet vanes, which washes the leaves and algae from the blades. Once that's finished, we open the valve fully to fill the box and restart the turbine.

RE Reliability

With a few lights and the stereo going, our household loads vary between 100 and 150 watts, depending on whether the fridge and/or freezer are running. If we use the microwave, toaster or electric tea jug, the load can briefly increase to 1,200 watts. With our microhydro system, these loads are no problem at all, and, within a few minutes, we're back to dumping the excess energy from the hydro plant. Unlike off-grid homes that rely solely on small PV systems, where *phantom loads* from TVs, stereos, computers and microwave clocks must be scrutinized, our hydro's continuous output means that we can just ignore them.

We recently added 492 watts of rooftop PV modules to our off-grid system. It's a cleaner, quieter backup to the microhydro system than an engine/generator and has turned out to be a wonderful complement. During instances of flooding or, more rarely, when the turbine has to be shut down for maintenance, we still have enough energy to run most of our common household loads.

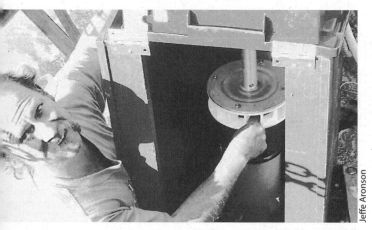

The new and improved cleanout procedure

The old, cold cleanout procedure

When the sun shines fully, the modules produce up to 450 watts. At the same time, the hydro is producing as much as 450 watts. When everything's humming, the systems produce as much as 13 kilowatt-hours daily! This is much more energy than we normally need, but having the two separate energy sources means that when the river is flooding or we're doing maintenance on the hydro, or conversely when the sun's been behind clouds for weeks, we still have plenty of energy.

With solar-made electricity, when the batteries are full, a charge controller cuts back the energy from the modules. But with microhydro, energy production does not stop — as long as water is flowing through the turbine. Switching off the electricity between the turbine and the battery will cause the turbine voltage to go too high. This scenario can create big problems, so those electrons have to go somewhere. When our batteries are full, a diversion controller shunts the excess hydro power to an air-heating resistance *dump load*, dispersing it as waste heat, and keeps the turbine electrically loaded and running at the proper revolutions per minute.

Have a Cold One

We are quite pleased with the way the systems are working. No longer slaves to our turbine, we can leave for days without worrying about food in the freezer thawing. During floods or long bouts of clouds and rain, we don't have to hassle with a backup generator. Even in rare instances when the river is in flood and there are also several days of heavy cloud cover, we employ energy-conscious practices to make the energy stored in our battery last for two or three days.

Hydro and PV Tech Specs

Overview

System type: off-grid, battery-based micro-hydroelectric with PV

System location: Victoria, Australia

Site head: six feet

Hydro resource: 30 gallons per second, dry season; 50 gallons per second, wet season

Hydro production: 252 kilowatt-hours per month average, dry season; 291 kilowatt-hours per month average, wet season

Solar resource: 4.7 average daily peak sun-hours

PV system production: 45 kilowatt-hours per month average

Hydro Turbine

Turbine: Energy Systems and Design LH1000, low head propeller

Runner diameter: five inches

Alternator: brushless permanent magnet

Rated peak power output: one kilowatt

Hydro — Balance of System

Hydro turbine controller: AERL Maximizer

Dump load: two 400 watt heat coils (resistors)

Inverter: Solar Energy Australia 2500, 24 VDC nominal input, 220 VAC, 50 hertz sine wave output

Circuit protection: 30 ampere breaker

System performance metering: Plasmatronics PL20

Photovoltaics

Modules: Sharp, ND-L3EJE, 123 watts STC, 17.2 Vmp, 12 VDC nominal

Array: 2 two-module series strings, 492 watts STC total, 34.4 Vmp, 24 VDC nominal

Array disconnect: 40 ampere breaker

Array installation: homemade steel mount installed on north-facing roof, 48° tilt

PV — Balance of System

Charge controller: Plasmatronics PL40, 40 ampere, PWM, 12-48 VDC nominal input and output voltage

Inverter: Solar Energy Australia 2500, 2.5 kilowatt, 24 VDC nominal input, 240 VAC, 50 hertz output

System performance metering: Plasmatronics PL20

Energy Storage

Batteries: Lucent 1AVR 2/85-75L, two-volt nominal, 1,200 ampere-hours, sealed, valve-regulated lead-acid (used telephone company batteries)

Battery bank: one 12 battery string, 24 VDC nominal, 1,200 amp-hours total

---

There are still some projects ahead. We'll install a more permanent, metal turbine box, with a discreet cover to protect the turbine from flood debris and to hide the equipment from hikers, anglers and our view. We'll be using some of our newfound time to plan a better pre-screening system to prevent debris or fish from working their way to the turbine box. Finally, we have a water heater coil awaiting installation in a hot water tank, so the ex-cess energy our system makes can be put to good use. After these upgrades, we'll take a long break and have that ice-cold beer, compliments of our fully functional RE system![1]

Editor's note: This article first appeared in *Home Power* Issue #122 (December 2007/January 2008) and is reprinted with permission of the author.

240 VAC Direct Drive Hydro

JOHN HERMANS

For seven years, my family and I have been deriving 100% of our electrical energy needs from a 240 VAC alternator driven by a low head, axial flow turbine. During this time and the years leading to its development and installation, I read many articles on microhydro electrical systems.

It is apparent that in recent years the number of microhydro installations have increased. This is due to their higher efficiency of operation and the availability of smaller, less costly units. These are primarily designed for constant battery charging. I am always intrigued to read of an installation that has a pure 240 VAC output, but it is quite rare to find such an article, thus I am prompted to write this one.

Site Selection

Of the hundreds of existing microhydro installations along the east coast of Australia, very few would readily adapt to a 240 VAC output due to the small volume of water on which they operate. But, there are a large number of potential sites that are suitable for 240 VAC which have low to medium head (3 to 20 feet) and flow rates which are considerably higher. This situation usually requires a river frontage property as opposed to a creek. It is the Nicholson River of East Gippsland, Victoria, Australia which provides us with just that: a low head (four feet) but high volume (26 to 528 gallons per second).

The Weir

Perhaps the only real obstacle to the construction of a similar system is the ability to build a water impermeable concrete and rock wall. This wall is required to allow water to enter the penstock (the piping to the turbine). It should not only resist flood damage, but should not initiate erosion problems in times of flood. Its potential to be undermined by water under pressure must be avoided. Such a site should preferably have a monolithic rock base for the majority of its length. The use of low level walls across rivers gives rise to minimal ecological impact. It is evident from my own installation with a small fish ladder that aquatic life is still free to move up and down the stream. This is most certainly not the case only half a mile upstream, where there is a 32 foot high concrete weir built to supply water to the nearby coastal towns. Low head installations are often subject to annual submergence by local flooding. In my case, this has minimal impact on the mechanical components. My only preparation for such a submergence is to remove and carry the alternator to high ground. This is a small price to pay for the high returns we receive from this system. Providing an installation with high flood resistance can be achieved if some basic principles are used.

To increase flood resistance:

- Use high strength concrete with water-proofing admixes at rock interfaces and include multiple drilled-in-place ½ inch rebar anchors.
- Use large diameter stainless steel anchor bolts with epoxy resins.
- Use heavy steel construction.
- Use pipe bracing.
- Use metal plate water and debris reflectors.
- Galvanize and epoxy paint steel that is frequently submerged.
- Position mechanical components out of line of flowing debris.

It is very possible for an entire system to be swept away. This threat has undoubtedly prevented many people from going ahead with their installations. On one occasion, I had 13 feet of flood water over the top of the alter-

One of the Pelton wheels

John Hermans

nator Power Tower for two days. Yet, my only chore after the waters subsided was to pump fresh grease into the bearings in order to displace water. After 20,000 hours, the running gear has been submerged at least 6 times, and the bearings have been changed only once.

Pumping Water

An additional benefit of my turbine is its use in driving a Grundfos multi-stage centrifugal pump. This pump delivers ⅓ of a gallon per second to a 58,000 gallon concrete tank situated 328 feet vertically above the river. This is achieved by taking the V belts off the alternator and flywheel and placing them on the pump.

Flywheel Use

The constant output of the four kilovolt-ampere alternator is relatively small (1.2 kilovolt-amperes), but is adequate for all our household and workshop requirements. (The exception is my welder, which is diesel powered. Putting out 250 amps at 40 volts does not come easily in many alternative energy systems). The use of a 66 pound, 15 inch diameter steel flywheel, spinning at 50 revolutions per second, is invaluable for starting induction motors up to 1.5 horsepower. Its stored energy is transferred smoothly to the alternator when sudden heavy loads are applied. This leads to minimal voltage and frequency drop, eliminating lighting flicker.

Design for Low Flow

The axial flow water turbine does not lend itself to throttling via reduction of the water flow through the turbine pipe. More than 50 gallons per second is required to run the turbine. In summer and other times of flows less than 50 gallons per second, a turbine shut-off valve is essential. A falling weir water level is a common problem in many installations and is overcome in my system in two ways. First, I use a bilge pump float switch in the weir water supply. This opens the control switch for the turbine valve when the water level falls approximately three inches. The control switch closes the turbine valve in the penstock, shutting the system down. When the water level is restored the switch closes to start the alternator spinning again. If the turbine takes 50 gallons to run constantly, but only 40 gallons was available, then the alternator-turbine would have a duty cycle of 80%, running 20 minutes on and 5 minutes off.

A preferred method of dealing with variable flow rates is to use a crossflow turbine which can be throttled by either a pivoting guide vane or a shutter arrangement across the water jet. This throttling ability, and the ease of home manufacture, makes this turbine a first choice for many low head installations. To counter low flows, we also use a small battery bank and inverter. Our TV, stereo and computer are all permanently wired to a 600 watt inverter. All other loads are automatically switched over to the inverter when the alternator switches off by either the float switch or manual controls. A slight flicker of the lights at night is all that is evident. In times of low river flow, most of my 1.2 kilovolt-amperes go into battery charging.

Frequency Control

The heart of this microhydro installation is the electronic stand-alone computer frequency control board. For four years, I had a very crude electronic load control which shunted unused power into either 400 or 800 watt dummy loads. This system was less than adequate on many occasions. My quest to find a more appropriate controller was not an easy one. After much searching, I chanced upon an electronic control board which not only works well, but leads the technology worldwide. I now have the first installation in Australia.

An Australian electronics engineer designed the frequency control board in conjunction with Appropriate Technology for Communities and the Environment (APACE) located at the University of Technology Sydney (UTS). Its development began when APACE, an Australian government aid program, began installing 240 VAC microhydro systems in the Pacific Islands. UTS has been using this control board as an educational tool for senior and post graduate students for over ten years. It has undergone thousands of hours of testing.

The circuit board itself is only six inches square. It uses a digital microprocessor incorporating a special algorithm to maintain alternator frequency to within one cycle per second on either side of 50 hertz. The software resides in ROM, and the unit is programmed to handle the high speed response required for low kilowatt output microhydro systems. For over three years, this unit has proved faultless in my system.

The control method maintains full load on the alternator by matching the power output to the load. This is accomplished by monitoring the frequency for variation and switching dummy loads.

In my case, the dummy loads are six incandescent lightbulb sets: 25, 50, 100, 200, 400, and 800 watts. The digital microprocessor determines which combination of the six loads to have on in order to maintain 50 hertz, and hence 240 volts. There are 64 combinations available, from 25 to 1,600 watts in 25 watt increments. Thus, if the house is using little of the generated power (say only 250 watts of night lighting), the controller will make up the difference. It will turn on the 50, 100 and 800 watt dummy loads to produce the total 1,200 watt load requirement. The load adjustments occur at the rate of ten per second, in 25 watt steps, which at times gives the appearance of

System Costs	
1 Concrete water tank	$4,000 (29%)
1 240 volt Pelton wheel/ alternator	$3,000 (22%)
1,475' Three-inch PVC pipe	$1,500 (10%)
2 Macron four kilovolt-ampere alternators	$1,300 (9%)
1,300' Paired cable	$1,200 (8%)
1 Grundfos centrifugal pump	$1,000 (7%)
1 Frequency controller	$1,000 (7%)
1 Power Tower	$600 (4%)
1 Concrete weir	$500 (3%)
1 12 volt Pelton wheel/alternator	$200 (1%)
Total	**A$14,300**

John Hermans

Hermans' Hydro System

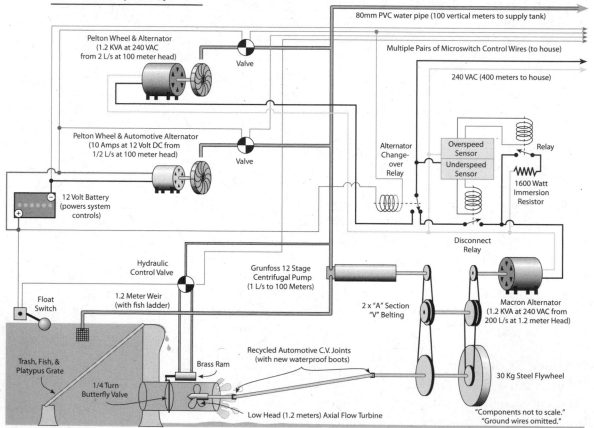

80mm PVC water pipe (100 vertical meters to supply tank)

Pelton Wheel & Alternator
(1.2 KVA at 240 VAC
from 2 L/s at 100 meter head)

Valve

Multiple Pairs of Microswitch Control Wires (to house)

240 VAC (400 meters to house)

Pelton Wheel & Automotive Alternator
(10 Amps at 12 Volt DC from
1/2 L/s at 100 meter head)

Valve

Alternator Change-over Relay

Overspeed Sensor

Underspeed Sensor

Relay

1600 Watt Immersion Resistor

12 Volt Battery
(powers system controls)

Disconnect Relay

Hydraulic Control Valve

Grunfoss 12 Stage Centrifugal Pump
(1 L/s to 100 Meters)

2 x "A" Section "V" Belting

Macron Alternator
(1.2 KVA at 240 VAC from
200 L/s at 1.2 meter Head)

Float Switch

1.2 Meter Weir
(with fish ladder)

Recycled Automotive C.V. Joints
(with new waterproof boots)

30 Kg Steel Flywheel

Trash, Fish, & Platypus Grate

1/4 Turn Butterfly Valve

Brass Ram

Low Head (1.2 meters) Axial Flow Turbine

"Components not to scale."
"Ground wires omitted."

a disco going on at the rear of my garage. The lightbulbs still manage a few years of use.

Damage Control

Included with my control system is a separate circuit board which is mounted close to the alternator. In case of control board or power line failure, it senses potentially damaging under and over frequency. It then responds by shutting down the turbine-alternator unit.[1]

Editor's note: This article first appeared in *Home Power* Issue #65 (June/July 1998) and is reprinted with permission of the author.

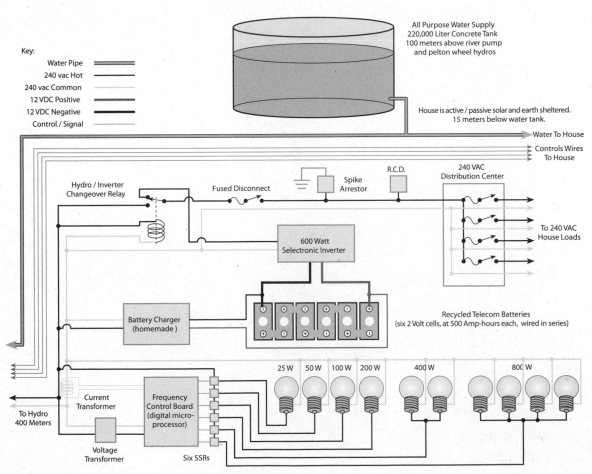

Key:
- Water Pipe
- 240 vac Hot
- 240 vac Common
- 12 VDC Positive
- 12 VDC Negative
- Control. / Signal

All Purpose Water Supply
220,000 Liter Concrete Tank
100 meters above river pump
and pelton wheel hydros

House is active / passive solar and earth sheltered.
15 meters below water tank.

Water To House

Controls Wires
To House

Hydro / Inverter
Changeover Relay

Fused Disconnect

Spike
Arrestor

R.C.D.

240 VAC
Distribution Center

To 240 VAC
House Loads

600 Watt
Selectronic Inverter

Battery Charger
(homemade)

Recycled Telecom Batteries
(six 2 Volt cells, at 500 Amp-hours each, wired in series)

25 W   50 W   100 W   200 W   400 W   800 W

To Hydro
400 Meters

Current
Transformer

Frequency
Control Board
(digital micro-
processor)

Voltage
Transformer

Six SSRs

Annex 4—Picohydro Pilots
Established in Ecuador

The following case studies are the results of a World Bank sponsored project called Energy Sector Management Assistance Programme (ESMAP) Ecuador.

Note the many differences between these systems and those from wealthier sources. These are primarily small AC units that produce entirely usable power, even at low outputs. Some use multiple turbines on the same water supply to provide for more than one household, thus neatly dealing with potential conflict.

Also of interest is that these life-changing turbines are much less expensive than their North American alternatives and yet contribute mightily to the reduction of pollution, and more. People who make US$2 a day spend a big fraction of that on kerosene for lighting. These small systems have beneficial impacts far in excess of their small size and expense.

The Guaslan Training Centre

The hydro resource for the Guaslan Training Centre comes from an irrigation channel out of the river Cebadas. The channel is three feet wide and 18 inches deep and has a measured flow of 300 gallons per minute and a natural

---

Guaslan Training Centre
Unit number: 6123

Province: Chimborazo
Place: Ciudad de Riobamba
Altitude: 8,940 feet
Coordinates: S01.667170 W078.634638
Responsible: Centro de Capacitación y Granja
 Integral "GUASLAN"
Source of water: Irrigation channel
Uses: This is a demonstration unit that powers
 three 20 watt lamps in the main hall of
 the training institution.
Flow: 950 gallons per minute
Head: Five feet
Distance to user: 262 feet

Unit Number: 6550

Province: Chimborazo
Place: Ciudad de Riobamba
Altitude: 8,940 feet
Coordinates: S01.667170 W078.634638
Responsible: Centro de capacitación y granja
 integral "GÜASLAN"
Source of Water: Irrigation channel
Uses: This is a demonstration unit that powers
 two 20 watt lamps and a radio for a small
 household inside the training institution.
Flow: 95 gallons per minute
Head: 20 feet
Distance to user: 300 yards

fall of about six feet near the training facilities. Civil works can provide a head of up to 164 feet.

It was decided to install two picohydro units (a low head and a high head) for demonstration purposes to disseminate the technology among visitors to the training center. The board of the center agreed to look after the units and to actively promote the technology.

Ozogoche Alto Community

The hydro resource for the Ozogoche Alto community comes from two rivers, the river Ozogoche and its tributary, the river Pichau-iña. The rivers both originate at the Ozogoche lagoons and are located 820 feet away from the community. The measured flow was 7100 gallons per minute.

To achieve the required head of five feet for the low head units, it was necessary to build a minimum infrastructure that consists mainly of channels to deviate water from the main streams. There is also a site at which a head of 20 feet could be achieved with minimum additional work.

The decision was made to install six low head units plus a high head unit. Because the community is highly organized, the ownership of each unit and the number of houses or facilities each unit will power was decided. Priority was given to communal buildings such as the church, school, community center and craft workshops.

Las Caucheras Community

The hydro resource for the Las Caucheras community comes from a small stream originating as runoff from the forested hills in the area. Because the area is quite steep, heads of 20 feet were easily achieved.

For that reason five high head picohydro units plus three low head units were installed.

Zapallo Community

The Zapallo community's hydro resource comes from the Rio Yatunyacu, a tributary of the river Napo. The topography allows most of the houses to be very close to the river. The measured flow was 5,550 gallons per minute, and the drop of the river in its trajectory around the community is 49 feet.

To achieve the required head of five feet for the low head units, it was necessary to build diversion channels for each unit. To simplify the civil work needed and to have another installation model available for future assessment, it was decided to build two small dams, each about three feet high. The walls of the dams were built of stones and sandbags.

On the first dam, a 164 foot long side channel was built to provide a five foot head. Three units were installed at the end of this channel. A hundred yards below the first dam, a second one was built. The same construction material was used. After a 1,300 foot long side channel, four units were installed. The remaining unit was installed out of the side channel.

The interesting feature of these installations is that they require additional maintenance because the side channels have to be looked after, and the dams have to be maintained regularly.

Alto Tena Community

The Alto Tena community's hydro resource comes from a small stream created by runoff from the forested hills in the area. Two of the units were installed on the Rio Pashishi Chico

Ozogoche Alto Community

Unit Number: 6128

Altitude: 11,318 feet
Coordinates: S02.252460 W078.601357
Responsible: Segundo Quishpe
Source of Water: Rio Pichauiña
Uses: The unit powers four households with a
20 watt lamp and a radio each.
Flow: 550 gallons per minute
Head: Five feet
Distance to user: 275 yards

Unit Number: 6350

Altitude: 11,318 feet
Responsible: Jose Bejarano
Source of Water: Rio Pichauiña
Uses: The unit powers four households with a
20 watt lamp and a radio each.
Flow: 550 gallons per minute
Head: Five feet
Distance to user: 250 yards

Unit Number: 6557

Altitude: 11,318 feet
Coordinates: S02.252460 W078.601357
Responsible: Communal church
Source of Water: Rio Ozogoche
Uses: The unit powers the main hall of the
church (two lamps and a small sound
system), its kitchen (one 20 watt lamp), and
two households with a 20 watt lamp and a
radio each.
Flow: 95 gallons per minute
Head: 20 feet
Distance to user: 220 yards

Unit Number: 6359

Altitude: 11,318 feet
Coordinates: S02.252460 W078.601357
Responsible: Jose Antonio Bejarano
Source of Water: Rio Ozogoche
Uses: The unit powers three households (one
20 watt lamp and a radio each) and a
cheese factory (one 20 watt lamp and a
radio).
Flow: 550 gallons per minute
Head: Five feet
Distance to user: 300 yards

Unit Number: 6134

Altitude: 11,318 feet
Coordinates: S02.252460 W078.601357
Responsible: Manuel Quishpe
Source of Water: Rio Ozogoche
Uses: The unit powers four households with a
20 watt lamp and a radio each.
Flow: 550 gallons per minute
Head: Five feet
Distance to user: 240 yards

Unit Number: 6127

Altitude: 11,318 feet
Coordinates: S02.252460 W078.601357
Responsible: Feliciano Bejarano
Source of Water: Rio Ozogoche
Uses: The unit powers four households with a
20 watt lamp and a radio each.
Flow: 550 gallons per minute
Head: Five feet
Distance to user: 260 yards

Unit Number: 6357

Altitude: 11,318 feet
Coordinates: S02.252460 W078.601357
Responsible: School and community center
Source of Water: Rio Pichauiña
Uses: The unit powers the school (two 20 watt
lamps), the teacher's house (20 watt lamp
and a radio), a workshop (two 20 watt
lamps) and a community center (two 20
watt lamps and a small sound system).
Flow: 550 gallons per minute
Head: Five feet
Distance to user: 275 yards

Las Caucheras Community
Unit Number: 6138

Place: Rio Arenillas
Altitude: 6,227 feet
Coordinates: S00 33 36.8 W077 52 29.6
Responsible: Luis Vitteri
Source of Water: Rio Arenillas
Uses: The unit powers a household with three 20 watt lamps and a radio.
Flow: 550 gallons per minute
Head: Five feet
Distance to user: 100 feet

Unit Number: 6352

Place: Rio Arenillas
Altitude: 11,318 feet
Coordinates: S00 33 36.8 W077 52 29.6
Responsible: Nancy Erazo
Source of Water: Rio Arenillas
Uses: The unit provides electricity for a household.
Flow: 550 gallons per minute
Head: Five feet
Distance to user: 853 feet

Unit Number: 6551

Place: Las Caucheras
Altitude: 7,234 feet
Coordinates: S00 36 44.7 W077 53 54.0
Responsible: Marco Silva
Source of Water: Spring
Uses: The unit powers two households; one is an ecolodge. In total there are 11 lamps (15 watts each), a radio and a small sound system.
Flow: 95 gallons per minute
Head: 21 feet
Distance to user: 164 feet

Unit Number: 6125

Place: Las Caucheras
Altitude: 6,276 feet
Coordinates: S00 33 42 W077 52 16.5
Responsible: Luis Alberto Vega
Source of Water: Spring
Uses: The unit provides power for a house with six 20 watt lamps, a radio and an electric fence (18 watts).
Flow: 550 gallons per minute
Head: Five feet
Distance to user: 295 feet

Unit Number: 6553

Place: Las Caucheras
Altitude: 7,276 feet
Coordinates: S00 39 36 W077 55 06
Responsible: Luis Aguilar
Source of Water: Spring
Uses: The unit powers seven 15 watt lamps, a television set (90 watts), a VHS unit (40 watts) and a sound system (100 watts).
Flow: 95 gallons per minute
Head: 39 feet
Distance to user: 100 feet

Unit Number: 6559

Place: Las Caucheras
Altitude: 7,303 feet
Coordinates: S0037 03.3 W077 52 16.5
Responsible: Edilberto Ayovi
Source of Water: Spring
Uses: The unit powers a household with two 20 watt lamps and a radio.
Flow: 95 gallons per minute
Head: 26 feet
Distance to user: 131 feet

Unit Number: 6555

Place: Las Caucheras
Altitude: 7,001 feet
Coordinates: S00 31 22.1 W077 52 67.8
Responsible: Samuel Llulluma
Source of Water: Spring
Uses: The unit powers a household with two lamps (20 watts each) and a radio.
Flow: 95 gallons per minute
Head: 26 feet
Distance to user: 500 feet

Zapallo Community
Unit Number: 6131

Place: Zapallo
Altitude: 1,673 feet
Coordinates: S01 04 50 W077 52 50
Responsible: Venancio Shiguango
Source of Water: Rio Yatunyacu
Uses: The unit provides power for a household with six 20 watt lamps; two sockets are available.
Flow: 550 gallons per minute
Head: Five feet
Distance to user: 131 feet

Unit Numbers: 6356, 6341 and 6122

Place: Zapallo
Altitude: 1,673 feet
Coordinates: S01 04 50 W077 52 50
Responsible: Elena Tapuy Clemente Andi Ines Tapuy
Source of Water: Rio Yatunyacu
Uses: Each unit powers a household with three 20 watt lamps and a radio.
Flow: 1,660 gallons per minute
Head: Five feet
Distance to user: 200 yards, 65 yards and 110 yards

Unit Numbers: 6136, 6355, 6130 and 6132

Place: Zapallo
Altitude: 1,673 feet
Coordinates: S01 04 50 W077 52 50
Responsible: Pedro, Jose, Luis and Carlos Andi
Source of Water: Rio Yatunyacu
Uses: Each unit powers a household. Three of the households have two lamps (20 watts each) and a radio. Unit 6136 powers a household with five lamps, a television set and a radio.
Flow: 2,200 gallons per minute
Head: Five feet
Distance to user: 140 yards, 119 yards, 239 yards and 112 yards

Alto Tena Community
Unit Numbers: 6120 and 6353

Place: Alto Tena
Altitude: 1,981 feet
Coordinates: S00 56 29.7 W077 52 52.9
Responsible: Nicolas and Carlos Grefa
Source of Water: Rio Pashishi Chico
Uses: Each unit powers 2 households with two 20 watt lamps and a radio each.
Flow: 1,100 gallons per minute
Head: Five feet
Distance to user: 70 yards and 120 yards

Unit Number: 6121

Place: Alto Tena
Altitude: 1,981 feet
Coordinates: S00 56 29.7 W077 52 52.9
Responsible: Francisco Aguinda
Source of Water: Rio Pashishi Chico
Uses: The unit powers three households. In total, there are four 20 watt lamps and two radios.
Flow: 550 gallons per minute
Head: Five feet
Distance to user: 40 yards

Unit Number: 6351

Place: Alto Tena
Altitude: 1,981 feet
Coordinates: S00 56 29.7 W077 52 52.9
Responsible: Francisco Tapuy
Source of Water: Stream
Uses: The unit powers three households and a poultry breeding hut. In total, it powers six lamps (20 watts each), a radio and a television.
Flow: 550 gallons per minute
Head: Five feet
Distance to users: 50 yards

and the other two on streams that are its tributaries.

On the Rio Pashishi Chico, two units were installed side by side. That was done to avoid excessive civil works and to take advantage of a small dam (three feet high) built to secure a constant flow of water into the turbines. A 131 foot long diversion channel runs from the dam to the two units.

The other two units, installed on small streams that are tributaries of the Rio Pashishi Chico, were installed at sites where there was a natural five foot head.

One Dam On Its Way

CHESTER KALINOSKI, RON MACLEOD

Dear HP,

Please give us some guidance. We are going to install a hydroelectric system at our homestead in the SW Virginia mountains. We have a flow range of 115 to 200+ cubic feet per minute. We will build a six foot high dam with floodwaters flowing over a rock-covered portion of an adjacent field that slopes toward the creek. The opposite bank faces a granite rock cliff. Since the pressure pipe run only needs to be about 40 feet long to avoid all floodwaters, we plan to use all the flow and build a small fish race for our minnow-sized trout. Thus, I believe we could use two fixed-blade propeller turbines, one at 70% flow and one at 30% flow.

Being a retired construction management mechanical engineer, I'm confident about doing the site work. What I need from you is a list of propeller and crossflow turbine manufacturers and suppliers of batteries, inverters, generators and electronic governors. Please also recommend some in-depth books on the subject.

Chester Kalinoski
Roanoke, Virginia

Hello Chester,

It sounds as though you have an interesting site with good potential, and your background lends itself to the work ahead. Your plans to construct a six foot high dam will require permission from the local fisheries folks and other state agencies. As long as you don't interconnect with the grid, the Federal Power Act of 1934 shouldn't apply. I don't think you need to notify the feds. A few hints on dam design for microhydro: it often pays to include a *draw down gate* in the dam design. This lets you drain the pond and remove accumulated silt and gravel from behind the dam (with state draw down permits). It also helps with dam inspection and maintenance. I would also include the turbine intakes, control gates and trash racks in the dam itself. You will probably place the intakes at the end of the dam opposite the cliff face, depending on bedrock. It is a good idea to position the intakes 90° to the stream flow just upstream of the dam. A floating timber can be placed across the front of the intakes to skim the surface and guide floating trash toward the spillway. Some folks use a notch in the top of the dam about two feet wide and four inches deep at the downstream end of the skimmer. This carries trash over the dam and also helps to insure some downstream flow below the dam.

When designing your turbine intakes at the dam, be certain to select the area of the rack below the surface to maintain a maximum velocity of ½ foot per second. Assume that the vertical bar rack will block about half of the area when you do the computations. Also plan to submerge your penstock's (pipeline) intakes at least two feet in order to avoid vortices. These could suck air into the system. I would recommend at least a ten-inch PVC penstock on each unit. Regarding equipment: your plan to use two units that split the flow ⅓ and ⅔ is good conventional practice on larger sites where it is important to optimize annual output. Depending on your expected loads, I would recommend a simpler approach. Two smaller units totaling about 70% of maximum flow should be sufficient. One hundred and fifteen cubic feet per minute on six feet of head will develop around 600 watts of power with my eight-inch Francis turbine. This is enough power when combined with the correct battery bank and inverters to run a home quite comfortably. A second unit could be added to use winter and spring flows if more power is needed or if you plan to handle some minor heating loads. You may want to size the first unit smaller to match extremely low flows.

I would contact the local office of the US Geological Survey (USGS) and do some research on potential flow at your site. Concentrate on the lowest flows in late summer and fall. Size your smaller unit to accommodate these numbers, and then go larger with the second unit. Always consider leaving some instream flow below the dam to support aquatic life. I'm sure that the state can help you with this. Your fish ladder designs will also have to be approved. From what I read in your letter, I feel that your site is best suited to either a prop or Francis-type reaction turbine.

Regarding possible turbine suppliers: Peter Ruyter manufactures small prop turbines in Sweden. The company name is Cargo and Kraft. You can contact him by email.[1]

I am going to introduce my small scroll case Francis units at the Midwest Renewable Energy Fair in Wisconsin this June. My Neptune design matches your site requirements well. Combined with the special brushless alternator I supply, and all-stainless construction, it should give many years of trouble-free service. Please feel free to contact me.

Best Regards,
Ron MacLeod[2]

Editor's note: This article first appeared in *Home Power* Issue #77 (June/July 2000) and is reprinted with permission of the authors.

PART IV

Microhydro in Context— History, Tips, Critical Design Elements

Context consists of those details that convey the feeling of experience. Microhydro exists in a historical and political context as well as being about plumbing and wiring. This part includes something from this larger picture with "Review of *The Death of Ben Linder*" (Chapter 34), "Small Water Power Siting" (Chapter 33) and "Microhydro Power in the Nineties" (Chapter 35).

Included here are also a couple of indispensable tips concerning plumbing and wiring in "Microhydro Pipe Dilemma" (Chapter 38) and "Soft Starting Electric Motors" (Chapter 36). As we saw in Chapter 27, lack of proper screening can make all the difference in system operation, and so "Microhydro Intake Design" (Chapter 37) is a topic that does need more exposure.

This part opens by providing an easy to understand method of estimating water flows, which is usually considered the most difficult element of a potential site to communicate. *Height* and *run of pipe* are similar to other measurements commonly made, but when it comes to gallons per minute, few have any extensive experience in estimating the flow of water. Comparing an unknown flow of water with carefully selected photos of known flows is good enough to start with for many projects.

32

The Photo Comparison Method of Estimating Water Flow

SCOTT L. DAVIS

While it is easy to measure the length of a proposed system and easy enough to measure the vertical drop, providing a reasonable estimation of the flow of water available can be quite difficult. Here's another way to give a rough estimate of the flow of water: compare the unknown flow with these images of known flows. If your flow looks to be more than the flow in one of the pictures, and less than the flow in another, then you can guess that the flow rate is between the rate of the two pictures. While a rough method, it provides an essential aid to visualization, particularly in the beginning.

Editor's note: This chapter is excerpted from *Microhydro: Clean Power From Water*, New Society, 2003.

20 gpm

Roy Davis

34 gpm

Scott L. Davis

100 gpm

just under 400 gpm

over 500 gpm

33

Small Water Power Siting

PAUL CUNNINGHAM

There are small streams running over much of the countryside. Perhaps you are wondering if a brook in your area is suitable for developing into a power source. The following is intended to show the procedure I used in my case to arrive at solutions to various problems.

How Much Is Enough?

A small scale water power system requires a more specific site than either a wind or photovoltaic one. You do need to have some flowing water. On the other hand, it isn't necessary to have very much, or much pressure, and it doesn't have to be very close to the point of use. My situation will illustrate this. Here in the Canadian Maritimes it is difficult to go very far without finding some type of stream. I live in an area of rugged topography which enhances the water power potential. My house is located near a brook that most times of the year has a fairly low flow rate. There is normally little water in the stream above the house while water from springs which come to the surface steadily increase the flow as the water runs downhill.

One logical place for the intake and beginning of the pipeline is near my house. Although flow increases further downstream,

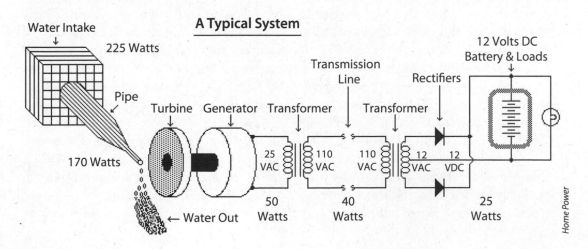

A Typical System

Water Intake
225 Watts
Pipe
170 Watts
Turbine Generator Transformer
25 VAC
50 Watts
← Water Out
Transmission Line
110 VAC
Transformer
110 VAC
40 Watts
Rectifiers
12 VAC 12 VDC
12 Volts DC Battery & Loads
25 Watts

Home Power

the slope decreases. Near the house the brook drops around eight feet for every 100 horizontal feet. So running a pipeline downstream 1,000 feet produces a combined drop or *head* of 75 feet. This looked like a reasonable place to start — although the site permits running a pipeline 3,000 feet before the brook meets another one running almost level.

One thousand feet of 1.5 inch polyethylene pipe was purchased (in 1978) and simply laid on the ground. A small screened box served as the intake and was set in the brook with a dam of earth and rocks sufficient to raise the water level about one foot. At this site, the maximum power will be produced at a flow rate of about 20 gallons per minute. This is the point where the dynamic (running or net) head is equal to two thirds of the static head. So there will be 50 feet of net head at the end of the pipe when the water is running with a suitable nozzle at the end.

Losses Within the Pipe

Any increase in flow will result in a decrease in power available due to increased pipe friction losses. Right away a third of the precious power potential is lost. At lower flow rates the pipe loss decreases which results in an increase in efficiency as flow decreases.

So why don't I use a larger pipe? Well, it costs more, and sometimes 20 gallons per minute is all there is in the brook. Also a larger pipe would aggravate the problem of freezing at low temperatures with no insulating snow cover. This is because the *residence time* (time water spends in the pipe) would increase with larger pipe. In my case, the water entering the pipe is (slightly) above freezing and cools

as it travels along (when temperatures are very low).

So why don't I bury it? Yes, that would be nice and hopefully I will when I can afford that and larger pipe too. It is a case of the shoemaker being inadequately shod as I content myself with the present system. Besides, it has spurred me on to other possibilities that we will look at later in future articles.

Nozzle Velocity

Back to the 20 gallons per minute at 50 foot head. A ⅜ inch diameter nozzle is about the right size for this, giving 19 gallons per minute. According to the spouting formula the velocity of a jet of water will be:

- Jet velocity equals the square root of two times the acceleration due to gravity times head in feet or
- 1.414 times 32.2 feet per second per second times 50 equals 56.7 feet per second.

Moving Water as Energy!

How much potential power is this? A US gallon of water weighs 8.34 pounds, and the flow is 19 gallons per minute; then 8.34 pounds per gallon times 19 gallons per minute equals 158 pounds per minute. Now, 158 pounds of water per minute falling 50 feet has 7,900 foot-pounds per minute of energy (simply multiply the factors). Conversion to horsepower is accomplished by division by 33,000; thus 7,900 divided by 33,000 equals .24 horsepower. Since 746 watts of energy is equivalent to one horsepower, .24 horsepower times 746 watts equals 179 watts of potential squirting out the nozzle. This means that the potential power was .36 horsepower or 269 watts before going through the pipe. Since nozzles tend to be very efficient not much loss is expected. But keep in mind that every time the energy goes through a change, power is lost. All right, how about a nine watt loss to make an even 170 watts?

This may appear a little sloppy. But you must realize that these systems do not have to be very precise — they are quite forgiving. Also many of the measurements are difficult to determine with high accuracy. So close approximations are sufficient.

Thus far things are reasonably straightforward — a pipeline with a nozzle at the end. Now what? Conventional practice would suggest some sort of impulse turbine such as a Pelton or Turgo. It would also be possible to use a reaction machine. It would have to resemble one of those spinning lawn sprinklers rather than say, a propeller type. This is because of the very small nozzle area. The impulse type looked easier to build.

Low Voltage DC Hydro

At this site it is necessary to send the power back upstream 1,000 feet to the house. I wanted to use 12 VDC and wanted some way to transmit the power other than the very large wire that would be required at this voltage. In the spring, when the flow in the brook was very high, various 12 VDC generators were operated with the pipeline ending near the house. But this could only be temporary, as ways of solving the transmission problem had to be discovered. Of course using wires wasn't the only possibility. I could always charge batteries downstream at the generator and then carry them up to the house. Or perhaps a reciprocating rod kept in tension could be used

to transmit the power? But all things considered, producing electricity at a voltage higher than 12 VDC looked the easiest.

Let's Raise the Voltage

I thought generating AC electricity at 60 hertz like regular commercial power would permit using standard transformers and make it easy to change the voltage. For this I bought a Virden Permabilt 120 VAC generator. This produces 1,200 watts rated output and 60 hertz at 3,600 revolutions per minute. These machines are reworked DC auto generators with rewound field, rotor with a slip ring and brush to carry the output.

An impulse turbine should have a surface speed of about half the jet velocity. So at 56 feet per second, a turbine wheel slightly less than two inches in pitch (hydraulic) diameter is required. This is a little on the small side, but I did make a Turgo wheel of this size so the rotational speed would be right for direct drive. Yes it's possible to use speed increasers with a larger turbine but I didn't think there was anything to gain — and only power to be lost. It turned out that the alternator would not generate 120 VAC at a low power level. The field required 10% of the rated 1,200 watts output to put out 120 VAC regardless of the load. Therefore a lower output voltage was necessary to properly balance the system. It was determined that under the site conditions an output of 50 watts at 24 to 25 volts was required to be in the correct ratio: 120 VAC divided by 10 amperes equals 24 VAC divided by 2 amperes or 48 watts.

Now you are probably wondering how come only 48 watts was being produced. Well that is what that combination of turbine and generator put out. And this isn't the end either. Next the juice went through a 25-110 volt transformer, through 1,000 feet of #18 AWG wire (two strands), another transformer down to 12 volts and then through rectifiers to give DC. In the end only 25 watts or about two amperes actually found its way to the battery.

This setup didn't last long enough to make many improvements. It was hard just keeping it alive. The alternator used only one slip ring. The other conductor was the bronze tail bearing! Both items had limited life under 24 hour service. Besides, the efficiency was low anyway.

A Functioning Higher Voltage System

I still needed a reasonable system…at least one with a longer life. In the next attempt a four-inch pitch Pelton turbine was cast in epoxy using a silicone rubber mold. This directly drove a car alternator with a rheostat in series with the field to adjust the output. Transformers (three) were connected to the three phase output to raise the voltage for transmission with the (now) three #18 AWG lines. Then a similar set of three transformers were used at the house to lower the voltage and a rectifier to make the DC conversion. About 50 watts was still generated (4 amperes at 12 volts), but more made it into the battery—about 3 amperes. The reason for this is the automotive alternators have more poles (12 Ford, 14 Delco) and generate at a higher frequency. This improves the efficiency of small transformers even though they are designed to work at 60 hertz. Now the system has an efficiency of around 21% (36 watts divided by 170 watts)

using the power available at the nozzle as the starting point.

What Can Be Done With 25 Watts?

Three amperes in a 12 VDC system doesn't sound like much. But this is sufficient to run the lights, a small fridge (Koolatron) and a tape player-radio. My house is small, and so are my needs. There was sometimes even extra power, and I could run Christmas lights or leave on things just to use the extra power.

At some point it occurred to me that I might generate more than electricity if I could produce turbines for others in a similar situation. Peltons were made first for sale. Originally these were made of epoxy and later of a high strength and abrasion resistant polyurethane. This endeavor busied me some, but it soon became apparent that to survive doing this sort of thing would mean producing complete generating units.

Turgos

Turgo turbines looked more reasonable than the Peltons for this, due to their greater flow-handling capability for a given size. Using a four-inch pitch diameter turbine wheel allowed as many as four one-inch diameter nozzles to be used. This resulted in a very versatile machine.

The first production models used automotive alternators (Delco) since they are inexpensive, dependable, available and most people wanted 12 VDC output. But these couldn't operate with heads of less than 20 feet or so. Also the efficiency of these alternators is in the 40–50% range, and I thought there was room for improvement.

Back at the R and D department, work was proceeding to develop a better machine. The Turgo turbines operate in the 60–70% efficiency range. These are made in reusable silicone rubber molds. But other tests showed there wasn't much to be gained by changing the shape of such a small wheel.

Permanent Magnet Generators

However, the generators used so far had efficiencies in the 50% range or less. They also had electric field coils which made for easy adjustment of the output but also took part of the output to operate. It looked like the use of a permanent magnet (PM) field would be a help and could make operation at very low heads feasible. Yes, DC motors with PM fields could be used as generators. But my experience with machines where brushes carried the full output was disappointing. Longevity was a problem — remember these are going to run 24 hours a day. If alternating current could be generated then transformers can be used to alter the voltage to suit the site.

It is well established that the most efficient generator type, especially in small sizes and at low speeds, is the PM-rotor alternator. Just like a bicycle generator. There is also nothing to wear out besides two ball bearings. That would be a feature and a half.

After a few tries, standard induction motors were used by keeping the stators and building new PM-rotors. This produced a machine capable of generating power with an efficiency of over 80%. Standard 60 hertz AC output was possible at 1,800 revolutions per minute for these four-pole machines. Experience suggested that frequencies of 50–400 hertz would operate standard transformers quite well. This, combined with the reconnectable output wiring, produced a machine able to generate almost any voltage.

Meanwhile Back At The Ranch

So how is it looking back at my site? Using the new PM-rotor alternator, about 100 watts of power is produced. This is an efficiency of 100 watts divided by 170 watts or about 59%. Dynamometer testing of the alternator shows it has an efficiency of 85% at this condition which means the turbine is running at 69%. Now 120 VAC is generated so no transformers are used at the generating site. The same transformer set used with the Delco installation is used at the battery end. About six amperes are delivered to the 12 volt battery. This gives an overall efficiency of 72 divided by 170 or 42% water to wire (water to battery?).

With this system appliances can be run directly off the alternator output as long as their requirement is less than the available power. This creates a hybrid setup that produces both 120 VAC at 60 hertz and 12 VDC. A future article will discuss how to deal with more difficult sites.[1]

Editor's note: This article first appeared in *Home Power* Issue #1 (November 1987) and is reprinted with permission of the author.

Review of
The Death of Ben Linder

CHRIS GREACEN AND ARNE JACOBSON

Ben Linder was a young renewable energy engineer, a unicyclist and a political activist working in rural Nicaragua to build micro-hydroelectric plants for the small villages of El Cuáand San José de Bocay. He was among those killed on April 28, 1987 when a crew making stream flow measurements for a planned microhydroelectric plant near San José de Bocay was ambushed by Contras, Ronald Reagan's "Freedom Fighters." In *The Death of Ben Linder: The Story of a North American in Sandinista Nicaragua*,[1] Joan Kruckewitt carefully pieces together the story of Ben's life during the nearly four years he lived and worked in Nicaragua. Joan's sources are interviews with friends and colleagues, as well as journals, letters and CIA documents.

The book is first and foremost about Ben — a committed idealist who believed in justice, peace and the potential for rural renewable energy projects to bring improved life to the rural poor. The book is also a powerful portrait of the suffering that US policy wrought on the people of Nicaragua.

Home Power readers who have worked (or dreamed of working) on renewable projects in developing countries will also find in *The Death of Ben Linder* a fascinating account of several years working on a 100 kilowatt village microhydropower project in the town of El Cuá. As the story unfolds, we learn what it took to get the project off the ground, to commission the plant in a war zone, to train local people to operate and maintain the plant, to electrify the surrounding communities and ultimately what electricity meant to people in El Cuá.[2]

Editor's note: This article first appeared in *Home Power* Issue #76 (April/May 2000) and is reprinted with permission of the authors.

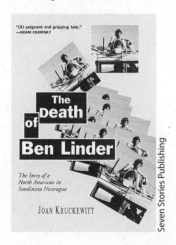

Microhydro Power in the Nineties

PAUL CUNNINGHAM AND BARBARA ATKINSON

Microhydro power was once the world's prominent source of mechanical power for manufacturing. Microhydro is making a comeback for electricity generation in homes. Increasing numbers of small hydro systems are being installed in remote sites in North America. There's also a growing market for microhydro electricity in developing countries. This article is a technical overview.

Microhydro power is gradually assuming the decentralized form it once had. Water power predates the use of electricity. At one time hydro power was employed on many sites in Europe and North America. It was primarily used to grind grain where water had a vertical drop of more than a few feet and sufficient flow. Less common, but of no less importance, was the use of hydro to provide shaft power for textile plants, sawmills and other manufacturing operations.

Over time thousands of small mills were replaced by centrally-generated electric power. Many major hydroelectric projects were developed using large dams, generating several megawatts of power. In many areas, hydroelectric power is still used on a small scale and is arguably the most cost-effective form of energy.

Renewable energy sources such as wind and solar are being scaled up from residential to electric utility size. In contrast, hydro power is being scaled down to residential size. The small machines are similar in most ways to the large ones except for their scale.

Siting

A hydro system is much more site-specific than a wind or photovoltaic (PV solar electric) system. A sufficient quantity of falling water must be available. The vertical distance the water falls is called *head* and is usually measured in feet, meters or units of pressure. The quantity of water is called *flow* and is measured in gallons per minute, cubic feet per second or liters per second. More head is usually better because the system uses less water and the equipment can be smaller. The turbine also runs at a higher speed. At very high heads, pipe pressure ratings and pipe joint integrity become problematic. Since power is the product of head and flow, more flow is required at lower head to generate the same power level.

More flow is better, even if not all of it is used, since more water can remain in the stream for environmental benefits. A simple equation estimates output power for a system with 53% efficiency, which is representative of most microhydro systems:

Net head in feet times flow in US gallons per minute divided by 10 equals output in watts

The size and/or type of system components may vary greatly from site to site. System capacity may be dictated by specific circumstances (e.g. water dries up in the summer). If insufficient potential is available to generate the power necessary to operate the average load, you must use appliances that are more energy-efficient and/or add other forms of generation equipment to the system. Hybrid wind/PV/hydro systems are very successful, and the energy sources complement each other.

The systems described here are called *run of river*; i.e. water not stored behind a dam (see *Small Power Siting*, Chapter 33, Ed.). Only an impoundment of sufficient size to direct the water into the pipeline is required. Power is generated at a constant rate; if not used, it is stored in batteries or sent to a shunt load. Therefore, there is little environmental impact since minimal water is used. There is also much less regulatory complication.

System Types

If electric heating loads are excluded, 300–400 watts of continuous output can power a typical North American house. This includes a refrigerator/freezer, washing machine, lights, entertainment and communication equipment, all

---

Definitions

Power equals the rate of doing work (watts or horsepower)

One watt equals one volt times one ampere

One horsepower equals 746 watts

1,000 watts consumed for one hour equals one kilowatt-hour (the unit used on utility bills)

Power is measured in watts and energy is measured in watt-hours.

Example: a 100 watt lightbulb uses power at the rate of 100 watts. During a period of 10 hours, it consumes 100 watts times 10 hours or 1,000 watt-hours or one kilowatt-hour of electricity

---

of standard efficiency. With energy-efficient appliances and lights and careful use management, it is possible to reduce the average demand to about 200 watts continuous.

Power can be supplied by a microhydro system in two ways. In a battery-based system, power is generated at a level equal to the average demand and stored in batteries. Batteries can supply power as needed at levels much higher than that generated and during times of low demand the excess can be stored. If enough energy is available from the water, an AC-direct system can generate power as alternating current (AC). This system typically requires a much higher power level than the battery-based system.

Battery-Based Systems

Most home power systems are battery-based. They require far less water than AC systems and are usually less expensive. Because the energy is stored in batteries, the generator can be shut down for servicing without interrupting

A Typical Battery-Based System

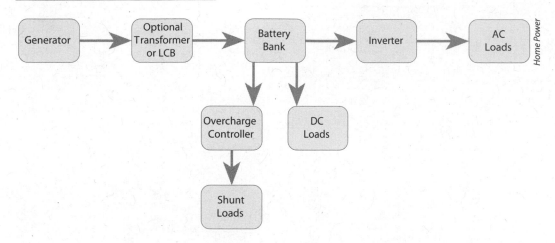

the power delivered to the loads. Since only the average load needs to be generated in this type of system, the pipeline, turbine, generator and other components can be much smaller than those in an AC system.

Very reliable inverters are available to convert DC battery power into AC output (120 volt, 60 hertz). These are used to power most or all home appliances. This makes it possible to have a system that is nearly indistinguishable from a house using utility power.

The input voltage to the batteries in a battery-based system commonly ranges from 12 to 48 volts DC. If the transmission distance is not great then 12 volts is often high enough. A 24 volt system is used if the power level or transmission distance is greater. If all of the loads are inverter-powered, the battery voltage is independent of the inverter output voltage, and voltages of 48 or 120 may be used to overcome long transmission distances. Although batteries and inverters can be specified for these voltages, it is common to convert the high voltage back down to 12 or 24 volts (bat-

tery voltage) using transformers or solid state converters. Articles on this subject appeared in *Home Power* Issues #17 and #28.

Wind or solar power sources can assist in power production because batteries are used. Also, DC loads (appliances or lights designed for DC) can be operated directly from the batteries. DC versions of many appliances are available, although they often cost more and are harder to find, and in some cases, quality and performance vary.

AC-Direct Systems

This is the system type used by utilities. It can also be used on a home power scale under the right conditions. In an AC system, there is no battery storage. This means that the generator must be capable of supplying the instantaneous demand, including the peak load. The most difficult load is the short-duration power surge drawn by induction motors found in refrigerators, freezers, washing machines, some power tools and other appliances. Even though the running load of an induction motor may

A Typical AC-Direct Microhydro System

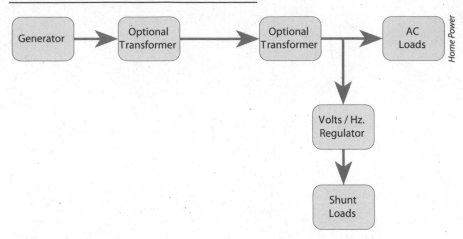

be only a few hundred watts, the starting load may be three to seven times this level or several kilowatts. Since other appliances may also be operating at the same time, a minimum power level of two to three kilowatts may be required for an AC system, depending on the nature of the loads.

In a typical AC system, an electronic controller keeps voltage and frequency within certain limits. The hydro's output is monitored, and any unused power is transferred to a *shunt load*, such as a hot water heater. The controller acts like an automatic dimmer switch that monitors the generator output frequency cycle by cycle and diverts power to the shunt load(s) in order to maintain a constant speed or load balance on the generator. There is almost always enough excess power from this type of system to heat domestic hot water and provide some, if not all, of a home's space heating. Examples of AC-direct systems are described in Part I and in Chapter 25 of this book. (Ed.)

System Components

An intake collects the water, and a pipeline delivers it to the turbine. The turbine converts the water's energy into mechanical shaft power. The turbine drives the generator which converts shaft power into electricity. In an AC system, this power goes directly to the loads. In a battery-based system, the power is stored in batteries, which feed the loads as needed. Controllers may be required to regulate the system.

Pipeline

Most hydro systems require a pipeline to feed water to the turbine. The exception is a propeller machine with an open intake. The water should pass first through a simple filter to block debris that may clog or damage the machine. The intake should be placed off to the side of the main water flow to protect it from the direct force of the water and debris during high flows.

It is important to use a pipeline of sufficiently large diameter to minimize friction losses from the moving water. When possible, the pipeline should be buried. This stabilizes the pipe and prevents critters from chewing it. Pipelines are usually made from PVC or polyethylene although metal or concrete pipes can also be used. The article on hydro system siting in Chapter 33 of this book (Ed.) describes pipe sizing

Turbines

Although traditional waterwheels of various types have been used for centuries, they aren't usually suitable for generating electricity. They are heavy, large and turn at low speeds. They require complex gearing to reach speeds to run an electric generator. They also have icing problems in cold climates. Water turbines rotate at higher speeds, are lighter and more compact. Turbines are more appropriate for electricity generation and are usually more efficient.

There are two basic kinds of turbines: impulse and reaction.

Impulse machines use a nozzle at the end of the pipeline that converts the water under pressure into a fast-moving jet. This jet is then directed at the turbine wheel (also called the runner), which is designed to convert as much

of the jet's kinetic energy as possible into shaft power. Common impulse turbines are Pelton, Turgo and crossflow.

In *reaction* turbines the energy of the water is converted from pressure to velocity within the guide vanes and the turbine wheel itself. Some lawn sprinklers are reaction turbines. They spin themselves around as a reaction to the action of the water squirting from the nozzles in the arms of the rotor. Examples of reaction turbines are propeller and Francis turbines.

Turgo Runner

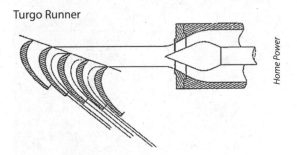

Home Power

Turbine Applications

In the family of impulse machines, the Pelton is used for the lowest flows and highest heads. The crossflow is used where flows are highest and heads are lowest. The Turgo is used for intermediate conditions. Propeller (reaction) turbines can operate on as little as two feet of head. A Turgo requires at least four feet, and a Pelton needs at least ten feet. These are only rough guidelines with overlap in applications.

The crossflow (impulse) turbine is the only machine that readily lends itself to user construction. They can be made in modular widths, and variable nozzles can be used.

Most developed sites now use impulse turbines. These turbines are very simple and

Pelton Runner

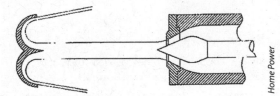

Home Power

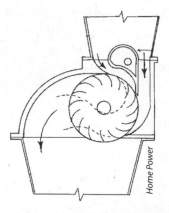

Home Power

relatively cheap. As the stream flow varies, water flow to the turbine can be easily controlled by changing nozzle sizes or by using adjustable nozzles. In contrast, most small reaction turbines cannot be adjusted to accommodate variable water flow. Those that are adjustable are very expensive because of the movable guide vanes and blades they require. If sufficient water is not available for full operation of a reaction machine, performance suffers greatly.

An advantage of reaction machines is that they can use the full head available at a site. An impulse turbine must be mounted above the tailwater level, and the effective head is measured down to the nozzle level. For the reaction turbine, the full available head is measured between the two water levels while the turbine can be mounted well above the level of the exiting water. This is possible because the *draft tube* used with the machine recovers some of the pressure head after the water exits the turbine. This cone-shaped tube converts the velocity of the flowing water into pressure as it is decelerated by the draft tube's increasing cross section. This creates suction on the underside of the runner.

Centrifugal pumps are sometimes used as practical substitutes for reaction turbines with good results. They can have high efficiency and are readily available (both new and used) at prices much lower than actual reaction turbines. However, it may be difficult to select the

Paul Cunningham, Energy Systems and Design

Paul Cunningham, Energy Systems and Design

Top: Crossflow turbine

Middle: This small turbine, the Waterbaby, can generate 40 watts from a garden hose.

Bottom: A Two Nozzle Turgo microhydro turbine

correct pump because data on its performance as a turbine are usually not available or are not straightforward.

One reason more reaction turbines are not in use is the lack of available machines in small sizes. There are many potential sites with two to ten feet of head and high flow that are not served by the market. An excellent article describing very low head propeller machines appeared in *Home Power* #23 and is included here as Chapter 24. (Ed.)

Generators

Most battery-based systems use an automotive alternator. If selected carefully and rewound when appropriate, the alternator can achieve very good performance. A rheostat can be installed in the field circuit to maximize the output. Rewound alternators can be used even in the 100–200 volt range.

For higher voltages (100–400 volts), an induction motor with the appropriate capacitance for excitation can be used as a generator. This will operate in a small battery charging system as well as in larger AC-direct systems of several kilowatts. An article describing induction generation appeared in *Home Power* Issue #3.

Another type of generator used with microhydro systems is the DC motor. Usually permanent magnet types are preferable. However, these have serious maintenance problems because the entire output passes through their carbon commutators and brushes.

Batteries

Lead-acid deep cycle batteries are usually used in hydro systems. Deep cycle batteries are designed to withstand repeated charge and discharge cycles typical in RE systems. In contrast, automotive (starting) batteries can tolerate only a fraction of these discharge cycles. A microhydro system requires only one to two days' storage. In contrast, PV or wind systems may require many days' storage capacity because the sun or wind may be unavailable for extended periods. Because the batteries in a hydro system rarely remain in a discharged state, they have a much longer life than those in other RE systems. Ideally, lead-acid batteries should not be discharged more than about half of their capacity. Alkaline batteries, such as nickel-iron and nickel-cadmium, can withstand complete discharge with no ill effects.

Controllers

Hydro systems with lead-acid batteries require protection from overcharge and overdischarge. Overcharge controllers redirect the power to an auxiliary or shunt load when the battery voltage reaches a certain level. This protects the generator from overspeed and overvoltage conditions. Over-discharge control involves disconnecting the load from the batteries when voltage falls below a certain level. Many inverters have this low voltage shut-off capability.

An ammeter in the hydro output circuit measures the current. A voltmeter reading battery voltage roughly indicates the state of charge. More sophisticated instruments are available, including amp-hour meters, which indicate charge level more accurately.

Conclusions

Despite the careful design needed to produce the best performance, a microhydro system isn't complicated. The system is not difficult to operate and maintain. Its lifespan is measured in decades. Microhydro power is almost always more cost-effective than any other form of renewable power.

Who should buy a microhydro system? In North America, microhydro is cost-effective for any off-grid site that has a suitable water resource, and even for some that are on-grid. Homeowners without utility power have three options: purchasing a renewable energy system, extending the utility transmission line or buying a gasoline or diesel generator. Transmission line extension can be expensive because its cost depends on distance and terrain. Even the initial cost of a hydro system may be lower. A gasoline generator may be cheaper to purchase but is expensive to operate and maintain. The life-cycle cost of the hydro system (3–25 cents per kilowatt-hour) is much lower than that of a generator (60–95 cents per kilowatt-hour). Once the hydro system is paid for, there's no monthly electricity bill and minimal maintenance costs. Since utility rates tend to rise, the value of the power increases, making your investment "inflation-proof." Notes to budding renewable energy enthusiasts: the future has potential if you use your head. There are many opportunities in this field for creative people with talents ranging from engineering to writing, if you're willing to find them and persevere. Remember what head, flow and love have in common: more is better![1]

Editor's note: This article first appeared in *Home Power* Issue #44 (December 1994/January 1995) and is reprinted with permission of the authors.

36

Soft Starting Electrical Motors

JIM FORGETTE

Important Caution

The permanent magnet motors used for converting washers and other machinery produce an usually strong, abrupt starting torque — more stress than many were designed to withstand. In the older washers with belt-drive and metal gears, a 12 or 24 VDC conversion seemed to handle this excessive starting torque, although over the years it will shorten the transmission's service life. The 115 VAC conversions can produce an even more severe, sudden start than 12 or 24 volts. Most newer washers use plastic type gears instead of metal. Direct-drive washers do not have a drive-belt to help dissipate this sudden jerk. The worst-case-scenario is a plastic-geared, direct-drive

washer converted to 115 VAC. This example could produce severe damage, in the short term, to the transmission gears. One detrimental effect from this strong start is that it prematurely degrades contact points of any relays (arcing). Another is that the permanent magnets in these motors can be prematurely weakened (demagnetized).

In certain industrial appliances using these motors, a *Soft Start* is included, which mellows this high amperage jerk start. Machinery you convert should also have this protection. I've researched this and put together a simple, effective soft start module using Rodan SG220, SG210 or SG260 thermistors. Two should be used for reversing-type washer/machinery

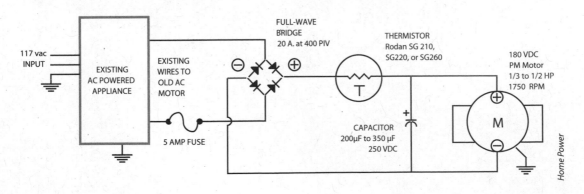

designs. You must know the motor horsepower rating and full load amps (common info on motor ID plates).

Summary

This circuit can reduce power consumption of many 115 VAC appliances and machinery in the ¼ to ½ horsepower range. Even though the new motor is DC, the appliance's power input is still 115 VAC. The existing AC supply/wiring can still be used. Improved efficiency will vary with application. Be sure to ground the DC motor's case to the frame of the appliance. Before removing the old AC motor, turn on the appliance and note direction that the motor turns. If the new DC motor runs in the opposite direction, simply reverse its DC power wires. If the new DC motor's revolutions per minute are low, then increase the value of the capacitor to 350 µF. Be sure to disconnect the power source from the appliance before working on it, otherwise you can get a nasty shock. Under load, the DC side of the circuit is about 145 VDC. If a newer motor's horsepower rating is the same or slightly higher than original AC motor, these circuit specifications will yield revolutions per minute very close to the original 1,750 rpm AC motor. This allows the original motor pulley/gear size and any drive belt to be reused in the conversion.

In a washing machine expect a three-fold increase in efficiency. The circuit can also be used in most conventional dryers, wringers, pumps, shop machinery, swamp-coolers, large attic fans, etc.[1]

Editor's note: This article first appeared in *Home Power* Issue #23 (June/July 1991) and is reprinted with permission of the author.

37

Microhydro Intake Design

JERRY OSTERMEIER

If you have a creek or stream on your property that drops along its course, a microhydro system could be in your future. With the right circumstances, harnessing microhydro electricity is cheaper and more constant than either photovoltaic or wind-powered electricity, partly because a source of flowing water is available 24/7 and not vulnerable to doldrums, clouds or the number of daylight hours.

While a microhydro system requires a more hands-on approach than PV, it is often simpler for the system owner than harnessing wind. It can also be a great do-it-yourself project — if you have the appropriate knowledge and skills. Supplying debris-free water to the hydro turbine is the first critical step in developing a low-maintenance hydro system. This article will introduce you to several methods of constructing an intake to do just that. Keep an eye out for additional articles that will cover penstock (the pipeline to the turbine) design and system wiring and transmission voltage considerations for high and medium head microhydro systems.

Creating a Diversion

If you follow a molecule of water through any high or medium head hydroelectric project, the first step is diverting it from its flowing source and into the penstock. A diversion can be a collection pond, a river-wide dam or even a pile of rocks that backs up the water in a creek enough to cover an intake. A diversion can be simple or elaborate, inexpensive or costly — but it needs to suit your application in a mechanical sense and also in an ecological one — without disturbing fish or their habitat.

This article does not cover some aspects of water diversion, like dam or pond building, since their construction is site-specific and can be quite complex, usually requiring professional design and engineering. They also often require permission and permits from government agencies, such as your state's fish and game department (in the US), and may need to include mitigation measures to protect fish and other wildlife.

Any diversion and intake needs to be robust enough to withstand the worst that winter has to offer — or it should be removable or easy to repair. Creeks can roll boulders or float trees that can damage your intake and diversion significantly. Some creeks flow evenly year-round, while others may trickle in the summer and flood in the winter. Because every site is a little different, whatever intake method you choose will need to be adapted to work at your particular location.

An important job of an intake is to screen out rocks and other debris, ¼ inch and larger, and anything that could lodge in the nozzles that direct the water stream onto the runner (the wheel in a turbine that is spun by the pressurized water). It also needs to keep out critters, like fish and other swimmers, and inhibit air bubbles from entering the pipe. For some turbine runners in high head installations, it's also best to filter out the *fines* (very small particles). Included in this article are the most common intakes used for this class of hydro, with pros and cons for each. Costs will depend upon its size and the choice of materials.

Hydro intakes

Simple Pipe with Screen

Benefits: Inexpensive
Drawbacks: Requires frequent cleaning
Cost: Low to moderate (US$30 to $400)

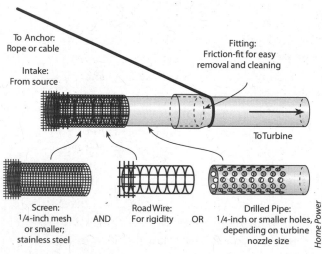

Simple pipe with screen

The simplest microhydro diversions are variations of a screen-covered pipe stuck in a creek. However, they require frequent cleaning. During the first rains of the season in some locations, twice-daily cleanings may be necessary to remove leaves and debris from the screen.

If placed in the stream's direct flow, the intake should be situated at least one foot underwater for pipe diameters up to four inches. This can create cleaning issues, especially during high-water periods. Although most folks aren't interested in wading into icy waters to clean their intakes, on small creeks these diversions can be reasonably nuisance-free most of the year, and the needed materials are inexpensive. The simple screen can be used in many different situations.

Open-Ended Pipe or Flume, with Settling Container

Benefits: Easy to clean; can be set up to filter out fine particles if needed; low maintenance
Drawbacks: Splashing and turbulence can cause air bubbles to enter the penstock; maintenance can be high if there are a lot of leaves
Cost: Low to moderate (US$30 to $400)

The next simplest diversion is an open pipe or flume that feeds a large vessel, such as a bathtub, 55 gallon drum or cistern. This approach offers easier access for cleaning than the pipe-with-screen method above. A ¼ inch wire mesh screen is secured over the container to catch larger debris. If needed, a finer mesh can be used under the ¼ inch screen to catch smaller particles. A cleanout can be placed at the bottom of the vessel to remove the fine sediment that makes its way through the screen.

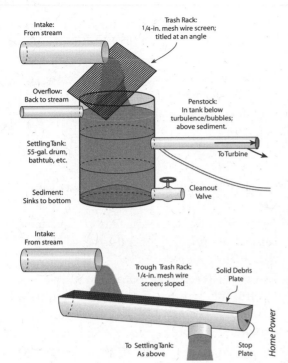

Open ended pipe or flume with settling container

In my system, a bathtub is located under the spillway of a small dam. A disadvantage to this bathtub design is that above flows of 300 gallons per minute, water coming over the spillway into the container tends to be turbulent and air bubbles can make their way into the penstock. This can result in lower turbine performance and also lead to early runner and bearing failure due to a water hammer effect. Using a deeper container, like a 55 gallon drum, usually fixes this problem. Maintenance is still medium to high in the fall because of leaves but is less during the rest of the year.

Screen Box
Benefits: Multiple screens catch debris of various sizes

Drawbacks: Moderately hard to clean because screens need to be removed from box
Cost: Moderate to high (US$100+)

Screen boxes are the second most common intake. They are generally constructed of concrete on-site, but can be fabricated of steel or plastic and brought in. Screen boxes are usually installed on the side of a small dam or slow-water area of a stream. They often have three screens, two valves and a cleanout. These work about as well as the bathtub diversion, but appear less lowbrow. They are harder to clean, especially during high-water events when the water level may be above the top of the box. If constructed without an outlet shutoff, debris can enter the pipe when the screens are removed for cleaning, which defeats using the screen box in the first place.

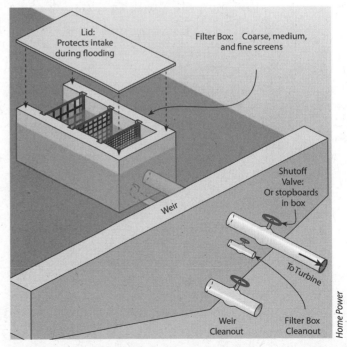

Screen box

Pond Bucket

Benefits: Simple to build

Drawbacks: Requires slow water with a stable level, plus access by boat or walkway; doesn't screen floating debris well

Cost: Low (US$30 to $50)

Another common method — the pond bucket — can be used close to the bank of a pond or slow-moving stream that has a fairly stable water level. With this approach, a pipe is run underground or through the dam to a screen. In a pond, the screen is often a five-gallon bucket with many holes drilled in the sides and top. Although this intake generally does its job in slow-moving and relatively clean water, large floating debris can strike the screen or bucket, dislodging it. And access to the intake for cleanout isn't very convenient: You'll either need a plank walkway or a

boat — or be prepared to swim. This diversion strategy is popular in areas with a lot of beaver activity because their dams won't impact the diversion. Pond bucket intakes can tolerate some heavy flows in the source, since pond levels can be controlled to some extent most of the time.

Culvert Tap

Benefits: Uses an existing culvert as a diversion

Drawbacks: Decreases capacity of culvert; can't accommodate higher flows easily; requires fabrication skills

Cost: Low to moderate (US$30 to $400)

This strategy taps into the bottom of a sloping culvert pipe near its lower end with a pipe adaptor to the penstock. A screen, angled low in the culvert at its upstream edge, in-

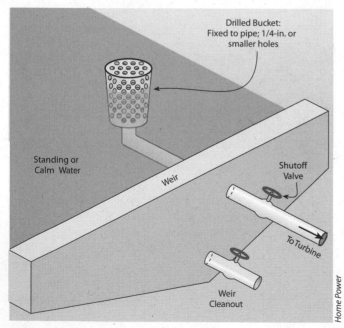

Pond bucket

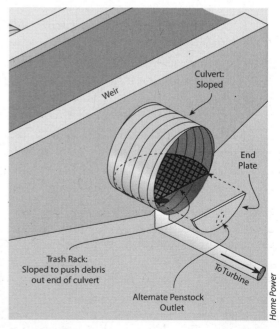

Culvert tap

creases in height relative to the bottom of the culvert at about three inches per foot. Because the slope of the culvert is still steeper than the slope of the screen within it, water can wash debris over the screen, making it somewhat self-cleaning. A wedge-shaped block at the end of the culvert and screen maintains the water level above the penstock adaptor.

Although this is a simple solution, be careful cutting or welding galvanized pipe — the fumes are toxic. Also be mindful of the maximum volume of water the pipe needs to carry during flood season. If a road washes out, you will be responsible for the damage.

Spillway with Coanda-Effect Screen
Benefits: Self-cleaning; easy to install; minimal ecological impact

Drawbacks: Expensive hardware; requires spillway

Cost: Moderate to high (US$900+)

If you can construct a spillway, a Coanda-effect shear screen, which uses the surface tension of water and triangular-shaped slots to suck water through without the debris (see *Home Power* Issue #71), can be an effective, nearly maintenance-free solution. In warm-water areas where algae tends to grow, it will be necessary to occasionally wipe off the screen.

Although a Coanda-effect screen is pricey, its stainless steel construction helps ensure its longevity — it should last forever, barring damage from trees or boulders. Many US state agencies now require this type of screen for microhydro applications because of its minimal

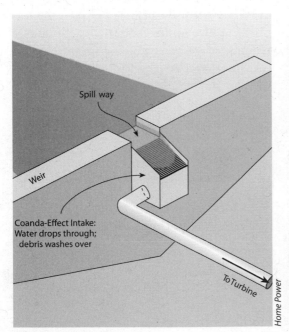

Spillway with Coanda-effect screen

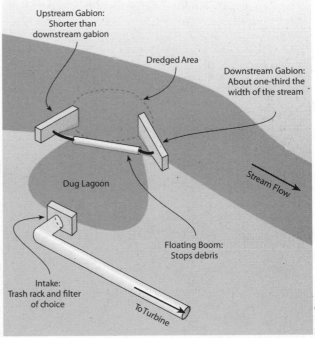

Slow water zone diversion

ecological impact—almost nothing makes it past the screen, including fish.

Slow-Water Zone Diversions

Benefits: Good for larger streams
Drawbacks: Complicated and time-consuming to build; potential for stream blowouts
Cost: High (US$1,500+)

These diversions are normally used on larger streams or rivers because of their expense, but can be used on smaller ones as well. I have built diversions in dry areas next to the stream and slightly below the existing water level, usually using concrete lagoons. When completed, simply breach the side of the creek to fill the new minipond. The important thing is not to redirect the stream flow out of its banks but just to pull water from it. I know of one case where the installer breached the side

of the creek and accidentally redirected the creek through a neighbor's house!

Gabion (a wire mesh cage filled with rocks) weirs can be extended from the bank to deflect the main stream flow enough to create a slow-moving (0.5 foot per second) water area to help settle debris. The deflector can also be made with rock and mortar. Gabion cages can be purchased flat from most irrigation supply stores, then assembled and filled on-site.

Siphon Intake

Benefits: Simple to install
Drawbacks: Loses prime, so restarting is often needed; still needs screening
Cost: Low (US$30 to $50)

A siphon intake over a diversion works by pulling water up a short rise, using suction from the downhill-flowing pipe below it. This kind of setup can work fine for pumping water, but tends to lose the prime in a hydro application. Because of this tendency, these intakes require a foot valve and a priming process (usually a hand-operated pump) to replace the trapped air with water when suction is lost.

Siphon Lift: Penstock rises out of streambed or over a rise

Siphon Drop: Weight of falling water to turbine overcomes small rise

Pump: Used to prime lift when suction is lost

To Turbine

Intake: Trash rack and filter of choice

Home Power

Siphon intake

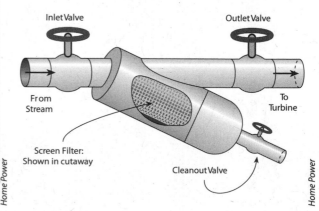

Inlet Valve

Outlet Valve

From Stream

To Turbine

Screen Filter: Shown in cutaway

Cleanout Valve

Home Power

In-line canister filter

They are used with a simple screen-type filter as part of the foot valve. Use this method if you must, but it is not advised — significant attention will be required, and you will eventually hate your microhydro system.

In-Line Canister Filter

Benefits: Ready-made to bolt in line with the penstock; easy to clean
Drawbacks: May require prescreening of fish and other objects; expensive hardware
Cost: High (US$1,500+)

In-line canister-type screens can work very well but still require prescreening to keep fish and larger objects out of the pipe. They usually have a cleanout port and bolt-up flanges for connection. I have used canisters made by Amiad Filtration Systems, but you should see what is available in your area in case you ever need parts. Water filters intended for households or drinking water will not work because they plug up too fast.

The Choice Is Yours

If you're lucky enough to have a creek that drops in elevation across your property, tapping into hydroelectricity is a great way to generate your own renewable energy. Properly installed systems will have minimal impact on the water source's ecosystem and its inhabitants. The range of intake and diversion types is as wide as a river. By reviewing the possibilities, you will find the right intake to fit your budget, the water source and your personal availability for cleaning and maintaining the system.[1]

Editor's note: This article first appeared in *Home Power* Issue #124 (April/May 2008) and is reprinted with permission of the author.

38

Microhydro Pipe Dilemma

BOB-O SCHULTZE

One microhydro pipe dilemma common with low head systems concerns air bubbles. Sometimes it is impossible or inconvenient to avoid having high spots in the penstock. Air bubbles can develop at these places and partially block off the flow of water, reducing the output. (Ed.)

...If you know the high point, get an air purge valve from a solar hot water dealer. When you're ready to install it, drain the pipeline — unless you need a shower. Drill and tap the valve into the PVC. Problem solved. In high head (pressure) systems, add a PVC tee and threaded reducer bushing, instead of tapping directly into the pipe.

Good luck,
Bob-O Schultze
(Electron Connection)

Editor's note: This article first appeared in *Home Power* Issue #114 (August/September 2006) and is reprinted with permission of the author.

The Hydro's Back

KATHLEEN JARSCHKE-SCHULTZE
(FROM HER COLUMN "HOME AND HEART")

I freely admit that I was kicked out of the glee club in the sixth grade because I couldn't carry a tune. I sing only for my family at our annual No Talent Show, for Bob-O and to my dog. So, with apologies to the Angels, the Chiffons and all those great girl groups of the 1950s and 1960s…

The Hydro's Back

[*Spoken in a soft, breathy voice*]
The creek went away and we hung around
And got our power from the wind and sun
And when the clouds came and no rain fell
We knew that things weren't very nice.

[*Begin singing to the tune of
"My Boyfriend's Back"*]

The Hydro's back and we're generating power,
Hey-la, hey-la, the hydro's back
The batteries are charging each
and every hour,
Hey-la, hey-la, the hydro's back.
The water's so high we're usin'
nozzle number two,
Hey-la, hey-la, the hydro's back
So look out now 'cause the amps
are coming through,

Hey-la, hey-la, the hydro's back.
Hey, the creek is finally flowing,
And the turbine's really going.
The creek's been gone for such a long time,
Hey-la, hey-la, the hydro's back
Now it's back and things'll be fine,
Hey-la, hey-la, the hydro's back.
Hey, every day I can do the laundry,
What to turn on next is my only quandary.
What made you think we always
needed blue skies?
Aah-ooh, aah-ooh
The wind turbine's quiet now,
but our amps are on the rise,
Aah-ooh.
Wait and see!
The Hydro's back,
it's gonna save our reputation,
Hey-la, hey-la, the hydro's back
So our Genny DC can take a long vacation,
Hey-la, hey-la, the hydro's back.
Yeah, my hydro's back…

The Drought Has Ended

We had a couple of intermittent inches of rain this fall. Then we had wonderful snow. Right here at Chateau Schultze, we got about six inches, and it stayed for a week. Our water year is at normal for this date, and the mountains all around us have visible snowcaps when the clouds clear away.

Our creek started slowly. Then you could begin to hear it whenever you went outdoors. It is a wonderful thing to live around free-flowing water. The water level in the creek rose steadily. After the anemic water flow all last winter, two years' worth of fallen leaves were finally flushed down the creek. We have had to take the walk through the meadow and up the creek to clean the hydro intake many times. This is not a hard job at all. Emma, the Airedale, loves going with us.

Our Hydro

The hydro unit is located in the creek just below our house. We use an Energy Systems and Design Stream Engine. It is about 80 feet from our battery bank and power room. Most of the 900 feet from the intake to the turbine is six-inch PVC pipe now. As it nears the turbine, there is five-inch and some four-inch PVC pipe.

It used to be all four-inch and three-inch pipe when we first started using hydropower in this creek. Every year while the creek is at its lowest (sometimes dry), Bob-O replaces sections of the smaller pipe at the upper end with larger pipe. This has resulted in more power being generated by the turbine.

The head or vertical fall of the water is about 32 feet. That is not going to change by any significant amount. So Bob-O improves what he can, when he can. After the creek had been running awhile, the water level rose enough to allow us to turn on the second nozzle of the turbine. That increased our generated power by eight amps at 24 VDC, almost doubling the turbine output at the time.

One Saturday, Bob-O built a new, larger base for the hydro unit. That gained us another half an amp. It allows the water to exit the unit more freely through a larger area as it flows back into the creek.

It is wonderful to have the hydro back. Bob-O and I look forward to it every year, and this year especially. The weather the last few years has been kind of wacky, and is getting drier overall...I am afraid to actually tally up how many projects I have either going on now or in my mind as having to be done come spring. Part of a birthday poem I once wrote for my sister said:

> Keep moving ahead, never look back,
> For work ahead, there's never a lack,
> Nose to the grindstone,
> shoulder to the wheel,
> We all ante up and take a chance on the deal.
> Hey, deal me in.[1]

Editor's note: This article first appeared in *Home Power* Issue #88 (April/May 2002) and is reprinted with permission of the author.

The Village or the House—
That is the Question

CHRIS GREACEN AND JEEVAN GOFF

There are two different perspectives on renewable energy rural electrification. One school of thought believes that village scale power production is the best option. Because of low cost, comparatively low tech manufacturing requirements and economies of scale, microhydroelectricity is most commonly chosen. Village microhydro plants are installed by a number of companies in Nepal, often using Nepalese-manufactured turbines and controllers. Projects usually involve working closely with a village to develop a managerial system, maintain the installation, collect fees from villagers and address conflicts that arise when villagers don't pay. Equipment usually consists of Pelton or crossflow turbines powering induction or synchronous generators. These generators vary in size from 100 watts up to hundreds of kilowatts. Electronic load controllers (ELCs) keep the voltage regulated by diverting excess electricity into resistive heating loads when households aren't using the hydro's full capacity (in the middle of the night, for example). In some installations there is no ELC, and the hydro plant is manually regulated by a man with a hand on the water valve and his eye on the voltmeter! In the past 20 years, some 300 microhydroelectric systems have been installed in Nepal. Another 1,000 produce mechanical power for grain milling and hulling or oil expelling.

A second school of thought focuses on decentralized home-based systems. This is the area where Lotus Energy has worked so far. It turned out to be much more difficult than expected to get an entire village to organize and follow through with hydroelectric projects. By contrast, household scale PV systems seem to be doing well. With stand-alone PV, each household gets its own system, each its own responsibility, and it alone enjoys the benefits. There's no social problem or headache! On the other hand, with individual systems, there's less opportunity for the village to work together on a common goal. In practice there is plenty of need for both PV and small hydro in Nepal, and the best technology for electrification depends on local geography and the goals and aspirations of villagers.[1]

Editor's note: This article first appeared in *Home Power* Issue #62 (December 1997/January 1998) and is reprinted with permission of the authors.

PART V

The Future—Bringing Market Solutions to Environmental Problems

Even if you believe that market forces are the root of all evil, microhydro equipment still needs to be sold and to be kept operating in order for its benefits to be felt. These benefits are remarkable. As well as being the most cost-effective renewable energy technology, every kilowatt-hour generated by water power saves 2.14 pounds of CO_2. This figure refers to power from utility-scale power plants and is likely low when compared to situations that use less efficient sources of energy such as kerosene. And, of course, this comparison does not take into account other pollutants generated by burning kerosene. The yellow flame of a lamp may make a pretty light, but it also means incomplete combustion, carbon monoxide and a host of other pollutants.

This part looks at possible contexts and futures for microhydro development. It is introduced by a retrospective, "Twenty Years of People Power" (Chapter 41), an historic look at how microhydro sites have fared over the years. Also discussed are predictions about the near future from that time, which is the future we now live in. How we have, and have not,

lived up to these predictions cannot help but inform us today.

Much microhydro resource remains undeveloped and unappreciated, wherever you look. Any opportunity to create value from the energy they produce makes systems just that much more possible. Thus the idea of "Energy Farming" (Chapter 42) and the example following it, "Practical Solar" (Chapter 43). Even a relatively small microhydro system, a few hundred continuous watts, can produce the same number of kilowatt-hours as these PV systems rated at thousands of peak watts. Microhydro projects make innovative ambitions like transportation and *practical* energy farming from renewable energy possible far beyond the reach of smaller PV systems. If a photovoltaic system can do it, likely your small hydro system will be more than enough. Chapter 43 shows the way. While this household uses the same amount of power as an average on-grid house, two-thirds of it goes to charge up their electric truck.

To reach audiences in the industrialized world, renewable energy still must overcome the myths and misperceptions that have grown up around it. These chapters offer many important examples of clean power in action to help dispel these myths.

It is, however, in the parts of the world where people make two dollars a day that microhydro makes its most profound impact. Whether you measure carbon footprint or increased productivity, the substitution of microhydro power for kerosene is without a doubt a remarkable improvement. Bringing this improvement into reality is the "The Story of PowerPal Microhydro" (Chapter 48).

Even nonprofits can use market forces to help solve environmental problems. As with renewable energy everywhere, financing is a powerful tool for overcoming the barrier of initial cost. Microfinance has been very successful in delivering credit service in just these areas. Small hydro units like PowerPal are priced well within the range of microfinance and should be an excellent match.

A project in development at Friends of Renewable Energy BC, *Streamworks*, matches these elements together (Chapter 49). This project offers benefits to the user (lights); benefits to their country (providing an opportunity to create greenhouse gas reductions in an easy to document fashion) and benefits to donors (to gain experience with renewable energy even if they don't have a site themselves, and without the full responsibility and cost of their own system).

Twenty Years of People Power

JOE SCHWARTZ AND IAN WOOFENDEN

There is more to *Home Power* than making electricity. It's easy for us to focus
on a piece of hardware: what it does, how it works, and how much it costs.
It's also easy to lose sight of the power that comes first — people power.
The will to do, and the power to accomplish what we will.

— Richard Perez, February 1988

When the first issue of *Home Power* magazine came off the press in 1987, it was greeted by a renewable energy (RE) landscape that was very different than it is today.

In the 1980s, only a few regions in the United States had experienced RE installers. Most first-generation systems were designed and built by off-grid homeowners who were tired of living with kerosene lamps or engine/generators, or often with no electricity at all. The idea of using RE equipment in home-scale applications was still revolutionary. Safety standards were in their infancy, and effective and durable system design approaches didn't have much of a track record.

As time went by, more and more systems were installed, but the RE community was fragmented. There was no Internet — many families living off-grid didn't even have phone service. Renewable energy information exchange happened at a snail's pace — or by the mail's pace.

Off-gridders Richard and Karen Perez had been selling and installing RE systems in their region for awhile, and were well aware of the information void. They dreamed of creating an interactive meeting place that would bring together the growing community of solar, wind and hydro users and facilitate an ongoing exchange of ideas to create better, safer and more productive renewable energy systems and equipment. Their vision became *Home Power* magazine, launched in 1987. With the first issue distributed free to 7,700 readers, the clean energy community came out of the woodwork. Hundreds of eager subscribers signed up each week, and renewable energy immediately became a technology bridge to common ground for back-to-the-land hippies, sun-toughened ranchers and assorted backwoods types of all persuasions.

The off-grid set of the 1970s and 1980s were energy pioneers who were willing to stick their necks out and put both their time and money on the line to try out a new generation of energy systems. Their motivation was the promise of cleaner, quieter and more cost-effective ways to generate electricity for rural properties that were beyond the reach of the utility lines.

Twenty years later, modern, grid-tied RE systems power everything from entire suburban subdivisions to downtown urban businesses. The early adopters of RE systems were fundamentally responsible for building the knowledge and experience base that helped pave the way for the flourishing renewable energy industry that is developing all over the globe.

Here's a look back — and a look forward — at just handful of the people, systems, homes and hardware that filled the pages of *Home Power* in its early years. The history it holds is a glimpse at the foundation of a movement that continues to change the way the world makes and uses energy.

Micro Mountain Power

People: Harry and Marlene Rakfeldt; Lisa and David Buttrey

Chapter 10 of this book — *Home Power* Issue #6 (August/September 1988)

Life throws a lot of changes our way. Sometimes homeowners move on, and their renewable energy system stays put. That was the case for off-gridders Harry and Marlene Rakfeldt.

Their rural adventure began when they bought remote property at 4,300 feet on the south side of Mount Ashland in southern Oregon in 1980. Although their land didn't have utility grid access, it had something better — two year-round creeks that dropped 300 feet down the steep and heavily forested mountainside.

In an interview for the August/September 1988 issue of *Home Power*, Harry pointed out that, "we like our creature comforts. We wanted our new home to be in all appearances the same as Dick and Jane's in the city." They tapped the renewable energy one of the creeks had to offer and installed a Harris hydro turbine, a bank of batteries and a Heart Interface inverter. Once the hydroelectric system was up and spinning, the Rakfeldts settled into what they felt was a pretty normal life up on the mountain: "We can curl up in front of the VCR for a double feature," said Harry.

While Harry and Marlene moved on from their mountain home, its remote location coupled with an existing renewable energy system attracted nature lovers Lisa and David Buttrey. David is a professional photographer, videographer and scuba diver; Lisa works with the Klamath Bird Observatory, helping to protect birds and their habitats. Lisa and David's professions are tied to the outdoors, and for them, living off-grid on a relatively remote, forested piece of land was a natural fit. They purchased the property in 1998.

When they moved onto the mountain, some components of Harry's 13-year-old microhydro system were beginning to show wear and tear, and newer, more advanced equipment was on the market. Local RE systems installer Bob-O Schultze modernized the system with eight new batteries, a new Harris hydro turbine and the latest in inverters at the time, a Trace SW4024. The new setup ran like a charm.

Nonetheless, the drought conditions that occurred during the early years of Lisa and David's tenure on the land meant relying on an engine/generator through much of the late fall, when stream flows weren't sufficient to power the house. They again enlisted the help of their local "power guru." In 2001, Bob-O installed 12

Siemens 90 watt solar-electric (photovoltaic; PV) modules off the south side of the home's lofty deck, and the Buttreys have had sufficient energy ever since. Many homesites have good access to the sun for solar electricity generation, but in comparison, properties with a hydro resource tend to be few and far between. "Our situation is enviable, certainly," says Lisa.

Northern Energy Independence

People: David and Mary Val Palumbo
Chapter 7 of this book — *Home Power* Issue #17 (June/July 1990)

Many RE industry pioneers got involved professionally after first powering up their own homes with renewables, and experiencing firsthand the benefits of clean, site-generated energy. David Palumbo was one of these early trailblazers. His off-grid homestead grabbed the cover spot on *Home Power*'s June/July 1990 issue.

Two houses and a combination horse barn/ workshop were built on the 100 acre property, each with its own battery-based power system that was charged primarily with PV. A propane-fueled engine/generator helped to keep the batteries charged up during Vermont's long and often sun-scarce winters. In 1987, a hydro system was installed and configured to charge the three separate power systems.

The original systems have remained largely intact. The first house they built on the property, now a rental, is still using a bank of 17-year-old nickel-cadmium batteries. The shop still gets some of its AC electricity from the original Heliotrope inverter, while a Studer sine wave inverter was added to power computers and sensitive electronics. More PV modules were added to each of the three systems over the years, along with a newer Harris permanent magnet hydro turbine. Combined, the site's current RE charging sources include 4.6 kilowatts of PV and a 400 watt microhydro system. David says, "Over the average year, we receive 90% of our electricity from renewable sources (50% PV and 40% hydro) and the remaining 10% from the propane generator."

After successfully developing his family's off-grid property in the mid-1980s, David founded Independent Power and Light, a highly respected RE design and installation business in Hyde Park, Vermont. "When we decided to make our home in the beautiful Green Mountains of northern Vermont, we had no idea where this new adventure would take us," says David. "Looking back at our decision...to produce our own electricity for our new homesite, I am amazed at how this one choice had such a profound effect on our lives."

RE Realists

People: Terry Kinzel and Sue Ellen Kingsley
Chapter 23 of this book — *Home Power* Issue #47 (June/July 1995)

Terry Kinzel and Sue Ellen Kingsley are practical people. Their no-nonsense approach to managing their small microhydro system has served them well, allowing them to wring as much energy as possible from the small stream that flows across their property.

In the summer of 1995, Terry and Sue Ellen shared their ten rules for living with a microhydro system in *Home Power*'s June/July issue. Their lessons steered prospective microhydro users toward realism — in design, equipment selection and installation, maintenance and system operation — no starry-eyed dreaming included.

They installed their system in 1991, tapping a stream that drops 16 feet with an average flow rate of about 75 gallons per minute. The energy in this falling water generates about three kilowatt-hours a day. The system has seen some modifications over the years. An unexpected flood damaged the hydro control equipment, and Terry and Sue Ellen opted to upgrade to a new turbine that gives them half again the energy. They also replaced their modified square wave inverter with a grid-capable sine wave model, and reconnected to the grid so they can sell their surplus energy and reduce the size of their battery bank. Terry says, "Our energy income exceeds our use by a small amount, and with the 24/7 production of the hydro and the grid as backup, our original batteries have never been discharged below 95% of their capacity." Conservation and efficiency have allowed Terry and Sue Ellen to live comfortably with a modest amount of renewable energy.

Terry and Sue Ellen are continuing to look for ways to diminish their use of fossil fuels and live more sustainably. They have installed a *living machine* wastewater treatment system that is growing a profusion of bougainvilleas and calla lilies. And they are putting more effort into education these days. Their home, extensive gardens and animals provide a setting for people to see an attractive life powered primarily by renewables. Sixteen years later, their microhydro system is still operational and providing much of the electricity for their home on Michigan's Upper Peninsula.[1]

Editor's note: This article first appeared in *Home Power* Issue #120 (August/September 2007) and is reprinted with permission of the authors.

Energy Farming

RICHARD PEREZ

Consider what would happen if homes came equipped with a 4,000 watt photovoltaic array. Most homes would instantly become net energy exporters. They would become energy farmers. Their "crop" would be sold to the local utility over the existing wires.

The technology to become an energy farmer exists today. The utilities wiring exists today. The sunshine exists today. What's stopping us from becoming energy farmers today? Only our inertia.

The barriers to energy farming are legal, financial and psychological. Energy has always been a commodity we bought from the power company. Our entire energy structure is based on centralized, utility-owned power production. They make and we buy it.

Times are changing. New legislation is favoring net billing for home-sized RE systems.

Utilities are being challenged over their monopoly on power production. And just plain folks are discovering the concept of energy farming.

It doesn't surprise me that technology is once again ahead of our ability to deal with it. We've got the hardware, but we're not sure what to do with it. Our energy establishment can't cope with the concept of energy farming. It challenges their hundred-year monopoly on electricity. Energy farming challenges our activism and dedication. Can we, as potential energy farmers, bring about this electrical transformation? You know we can.

Editor's note: This article first appeared in *Home Power* Issue #46 (April/May 1995) and is reprinted with permission of the author.

Practical Solar—A Californian Combines Net Metering with Time of Use Metering to Make PV Pay

PHILIPPE HABIB

I never wanted to be a power producer. I came to it when the advantages couldn't be ignored anymore. When grid power was reliable and cheap, there was no reason to spend a pile of money on a photovoltaic (PV) system. I thought that a PV system would not only never pay for itself, but that it would double my energy bill. Besides, the last thing I needed was another thing to take care of. You never see a magazine called *Utility Power User* on the magazine rack, and there's a good reason for that. Utility power just works; it doesn't require a support group to pass around hints and tips.

It wasn't that I didn't feel capable of designing or installing a system. I designed our house, and I was very involved in its construction. But with a busy life, I don't want to spend the little free time I have adjusting this or that or troubleshooting on a regular basis.

Thermal Efficiency

My wife and I kept with the pragmatist theme as we were designing and building the house. Energy use and efficiency were an important but not overriding part of every decision. All windows are double glazed, and all south or west-facing glass is low-E. All of the insulation exceeds the typical R-19 ceilings, R-13 walls and R-13 floors in our area—we have R-30, R-15 and R-19 respectively. All hot water and house heat comes from a very high efficiency Polaris model water heater made by American Water Heater. The tank of this 94% efficient

Half of the Habib family's utility-intertied PV system was paid for through California's buydown rebate program.

Philippe Habib

unit is made entirely of stainless steel. So in addition to fuel savings, I also avoid buying a new heater every few years. The heat system is hydronic, using cross-linked polyethylene pipes in the floor. In addition, two 4×8 foot solar thermal panels preheat water, which is stored in a 100 gallon tank for domestic use.

We also installed a masonry heater, made by Temp-Cast Enviroheat, for our main source of heat. It burns a very hot, low-polluting fire for a fairly short time and stores the heat in masonry walls a foot thick. The heat is let out slowly over the next 24 to 36 hours. Three to four fires a week is all it takes in our mild climate to keep the house warm.

Payback was the Point

We are not so committed to conservation that we will make big changes in our lifestyle or spend lots of money just to save energy. We built a practical house that uses less energy than other houses of its size because we chose to spend more upfront on features that have a long-term payback. But every feature is there because it has a payback, not because we

The utility-intertie Trace Sun Tie 2500 sine wave inverter (right) and the upstairs AC subpanel

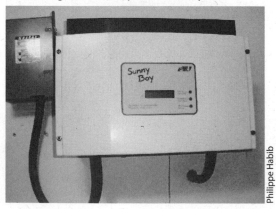

Philippe Habib

would do anything to raise efficiency, no matter the cost. With that attitude, I did a bit of research to confirm my negative biases about PV power. What I found really surprised me. With California's Emerging Renewables Buy-down Program 50% rebate, I could give my money to the utility every month, or I could use it to pay off a PV system. The cost would be the same — until the utility raised its prices.

In June of 2000, when I was first designing this system, the tripling of prices in San Diego was in the news. So I expected something similar to happen when my utility, Pacific Gas and Electric (PG&E), was permitted to raise its prices. I was impressed both by the long warranties on the PV modules and the fact that a lot of the systems installed thirty years ago are still running and still producing power. If done right, it looked like this could be an install-it-and-forget-about-it kind of deal.

Our monthly electrical use is about 900 kilowatt-hours. About 600 of that goes to charge my electric Ford Ranger pickup, and occurs at the off-peak rate between midnight and 7 AM. I measured how much space was available on my roof and priced components. It looked like a 2,000 to 2,500 watt system would be a good fit for both space and energy produced.

Sun Tie

At the time I was designing the system, Trace was about to introduce the Sun Tie (ST) series of inverters, which promised to lower the cost and ease the installation of a grid-tied system. Normally, I would be wary of buying a brand new product, but I figured that Trace had been in the inverter business long enough that a new product might not be a big risk.

As things turned out, the display on the unit I received was dead out of the box. It took about six weeks between my first contact with Trace and the replacement of the inverter. The second inverter's display only worked for the first few hours of each day. Eventually, I did some troubleshooting and found the problem. The display was in close proximity to the inverter's power components, and electrical noise was causing the display to fail. Wrapping it in a grounded copper envelope solved the problem.

This was more involvement than I wanted with a product, but I did know that I was buying an early release, and the process was kind of fun, anyway. Other than the problem with the display, the Sun Tie is a wonderful product. Getting it working consisted of nothing more than bolting it to the wall, and attaching the DC wires from the panels and the AC wires to the house electrical panel. All breakers and fuses are factory installed. There is nothing to set up or configure—it just works. In order to get my cost per installed watt down as much as possible, I went with an ST2500 inverter and 24 Kyocera 120 watt panels. The cost per watt of the ST inverter goes down as the capacity goes up. An ST2500 does not cost 2.5 times the price of an ST1000. I figured that once I got the lowest dollars per watt inverter I could find, the best value would be to load it up as fully as possible with PVs. That meant I wasn't paying for any inverter capacity that I wasn't using.

For the California buydown rebate, the calculations look like this: 24 120 watt PV modules have a PVUSA Test Conditions (PTC) value of 105.7 watts each, for a total of 2,537 watts. The 94% efficient inverter makes a system output of 2,385 watts total.

Habib System Costs	
24 Kyocera 120 watt panels	$11,880
Trace ST2500 inverter	$1,725
Wiring and mounts	$400
Service contract*	$225
Total	**$14,230**

Note: costs in US dollars.

* To bring the warranty to five years, as required for the buydown program.

Philippe Habib

I was ready to order the equipment from out of state to save the sales tax. But when I heard that I'd have to pay it anyway to get the buydown, I figured that I'd keep my money in state. I wound up making my purchase from Solar on Sale, who beat the out-of-state price anyway. The people there have been very knowledgeable and helpful. In all, the system cost me about US$6 per watt. I'll get half of that back from the buydown program.

I did all of the installation with the help of my friends Greg Stefancik and Dave Kucharczyc. Since I have a pretty complete metal shop in the basement, I made my own mounts using aluminum angle and stainless steel hardware.

TOU with Net Metering?

I did some research, and learned that my utility, Pacific Gas and Electric, was obligated to offer net metering (E-NET, the PG&E tariff schedule that deals with net metering). Now, that was interesting. As the owner of an electric car, I'm on the E-9 tariff schedule. This is a three-tier time-of-use (TOU) rate. The two lower priced tiers are offered in the winter, and the third, highest-priced tier is for the summer months during the afternoon.

The rate schedule is set up to discourage use during peak load time and to encourage you to charge your EV between midnight and 7 AM. The peak cost coincides with a PV system's peak production. I concluded that a PV system could pay for itself just by saving me from buying that expensive summer peak power.

I read the tariffs on PG&E's website. It wasn't clear to me how net metering worked for TOU customers. Would surplus generation be credited to me in kilowatt-hours spread out over my bill? Or would it be in dollars applied to my purchase of power at a lower rate later?

Bureaucratic Goose Chase

In getting the special rate for my truck charging, I learned that most of the people who answer the phones at the utility don't have the training to be of help with unusual questions. You have to work to find the person who can really answer the more complicated questions. For instance, within an hour of getting in touch with Efrain Ornelas, alternative vehicle program manager for PG&E, my EV account was set up. Prior to that, I'd had at least half a dozen conversations in a three-week period.

Figuring that the alternative vehicle program manager would know who I should talk to about E-NET, I called and asked him for help. He put me in touch with Harold Hirsh in the renewables department. According to him, the answer was that I'd be credited in neither power nor money. He said that PG&E did not offer net metering with TOU rates. I would have to switch to a non-TOU rate if I wanted net metering. Funny, I never saw anything that said I'd have to be on a particular tariff to get E-NET.

So I made a few phone calls to the California Energy Commission (CEC) to try to clear things up. No one there had the definitive answer, but one name kept being mentioned as the person who would know — Vince Schwent (now with Sacramento Municipal Utility District). I called him, and he said my take on it was correct. Every tariff was eligible for E-NET, and I should be credited in dollars, not power. This meant I could sell my summer afternoon excess at US$0.30 per kilowatt-hour, and buy that power back at night to charge my truck at only US$0.04 per kilowatt-hour!

I went back to PG&E, and they said that it was not technically feasible to do E-NET with TOU because the TOU meter was not capable of going backwards. The meter apparently treats all energy going into the grid as theft, and never allows the count to decrease.

Another conversation with the CEC pointed me to the Sacramento Municipal Utility District (SMUD). They have a comprehensive renewable energy program, and their meter shop had done a lot of meter testing. They found that General Electric makes a TOU meter that can accurately register backwards. This is their model KV.

During my next conversation with PG&E, I mentioned the GE model KV meter. Apparently there were lots of reasons why it wouldn't work for me. PG&E didn't use that meter and could not use it because of PUC regulations. They couldn't take SMUD's word that it accurately read in both directions. And the computer billing software wasn't set up for TOU and E-NET, so it wouldn't work anyway.

The tariff is pretty clear on one point — PG&E had three options:

1. Provide me with an appropriate meter.

2. Allow me to purchase such a meter.
3. Pay to have my house rewired to use dual meters.

Since they claimed that they couldn't do #1, I was prepared to do #2. In fact that was my preference, since I wouldn't have to make a lot of US$12 monthly meter rental payments to pay for a US$300 meter. The threat of option #3 seemed to be the way to get them to do one of the other two.

Further conversations with PG&E yielded more information. It turned out that they do use the KV meter, but only on some industrial accounts. So much for it not being in their inventory and unavailable to their meter shop.

Net Metering Solution

Eventually, I got connected with Phil Quadrini from PG&E's headquarters in San Francisco. He is the guy who knows the arcana of tariffs inside and out. He agreed that TOU and E-NET did not need to be mutually exclusive, and he figured out a way to make it all work.

I would get a GE model KV meter, which the meter shop would pre-load with 50,000 kilowatt-hours as the starting point reading. Since the meter can't show negative numbers, this was necessary so that the meter will not stop at zero if my surplus ever exceeds my use. At the end of the one year averaging period, PG&E would read the meter and I'd pay for any net use during the year.

Once that was settled, Phil put me in touch with Jerry Hutchinson, who sent me the application package to get me legally connected to the grid. I sent in my application to connect my PV equipment to the grid and waited. Six weeks later, I still hadn't received anything

back. I called to check the status of my paperwork.

The person who processes the applications told me I'd have to switch to a non-TOU rate to do E-NET. Apparently, the word had not filtered down. A few more phone calls referring back to my earlier conversations with Phil straightened that out. I now had a commitment from PG&E to combine TOU with E-NET.

It took a lot of perseverance to get the utility to go along, but everyone wins with this arrangement. This makes a PV system more affordable and therefore more feasible for homeowners. And for the utility, buying my surplus, even at about US$0.30 per kilowatt-hour, is cheaper than the high peak prices (which recently went above US$1.40 per kilowatt-hour) when demand gets high. And the overloaded transmission network does not have the additional burden of bringing me energy from far away.

Since getting this arrangement, I've learned that I'm not the only one who sees the benefit of it. California State Assembly member Fred Keeley authored AB-918 to clarify the issue of E-NET and TOU. This legislation (now California law) requires the utilities to offer them together, and to buy back power at the retail cost of that power at the time it is generated. So there is now a very clear mandate that it must be done this way.

Not Just for Tree Huggers

A system like mine shows that PV power is no longer just for tree huggers and those who live far from the grid. If this system did not promise to be reliable and cost-effective, I would not have installed it. Getting the E-NET and

Philippe Habib's Utility-Intertied PV System

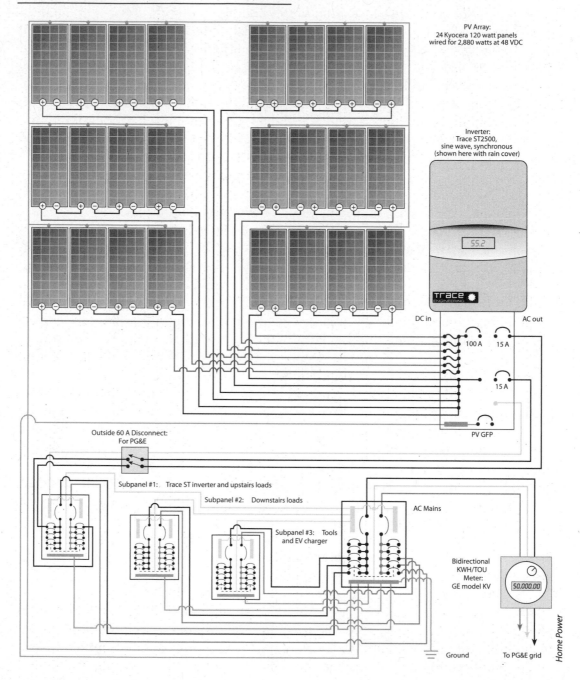

PV Array:
24 Kyocera 120 watt panels
wired for 2,880 watts at 48 VDC

Inverter:
Trace ST2500,
sine wave, synchronous
(shown here with rain cover)

55.2

Trace
ENGINEERING

DC in

100 A

15 A

AC out

15 A

PV GFP

Outside 60 A Disconnect:
For PG&E

Subpanel #1:   Trace ST inverter and upstairs loads

Subpanel #2:   Downstairs loads

Subpanel #3:   Tools
and EV charger

AC Mains

Bidirectional
KWH/TOU
Meter:
GE model KV

50,000.00

Ground

To PG&E grid

Home Power

TOU helps to make the payoff even more attractive, but the economics are there even without it. If you don't have an electric vehicle to qualify you for the E-9 rate, you can get E-7. Right now, the cheapest E-7 rate is US$0.085.

The E-7 TOU rate still allows you to sell your summer afternoon surplus at high rates and apply that money to your off-peak usage. You just don't get the ultra-low rate from midnight to 7 AM. Of course, this assumes that you have a summer afternoon surplus, and that you're not dipping into the grid to run an air conditioner.

Now that the system is installed and has been working for a few weeks, I can say that I'm very pleased with it. On a nice day in February, I can get about 10.5 kilowatt-hours of energy from the PV array, and on a partially overcast day, I might get six to eight kilowatt-hours. A couple of weeks ago, I got a bit more than 15 kilowatt-hours. Once summer starts, I would not be surprised to see more. The best part is knowing that no matter what happens, my cost for electricity is frozen. Not contributing to pollution is nice, but it was not my overriding reason.

RE Goes Mainstream

I'm reasonably handy, but had no prior knowledge about anything solar. And I put together a working and cost-effective PV system. This shows that the products have matured to the point where PV is not just for the isolated or the ideological anymore. Homeowners who live in suburbia can decide to install a PV system purely for financial reasons, and still make it work out.

I do realize that I owe a debt to the traditional readers of publications such as this one. You are the true believers who brought things along to this point. There may even come a day when *Home Power* magazine is as tough to find on the newsstand as *Utility Power User*...[1]

Editor's note: This article first appeared in *Home Power* Issue #83 (June/July 2001) and is reprinted with permission of the author.

Independent Power Providers—
Beyond Net Metering

In the last two issues of *Home Power* magazine there have been constructive critiques of current and pending net metering laws. In *Home Power* Issue #63, Jeff Klein's letter pointed out that most net metering laws were flawed because they did not include wind and hydro sources. Jeff also said that many grid-connected PV systems were installed by homeowners who did it for environmental or hobbyist reasons, not economics. *Home Power* #64's Power Politics column makes some of the same points and questions a pending change in California's net metering law that gives away any net excess power to the utility or provider.

A Second Look

Though it seems galling to give power away to the utility, this is not such a bad deal. Let's examine the economics. In addition to a zeroing out of net excess energy is a change from a monthly to a yearly accounting period. Monthly summer excess credit can be carried forward. Unless the array is oversized (unlikely), winter months will allow the excess credits to be used up. The customer-generator receives retail value for any monthly excess

power produced. The annualization period for net metered PV systems would best be timed to start in the spring. Under the old method, monthly accounting of excess power would only yield avoided cost (wholesale, about two cents per kilowatt-hour) and monthly settlement would, at best, be a couple of dollars.

It's Worth More

Should the RE generator get more than retail for clean power? Independent Power Providers (IPP) thinks so. In Europe, another approach called *rate based incentives* has been used, yielding owners of PV generated electricity around 40 cents per kilowatt-hour. Some of those programs were discussed a couple of years ago in this column. Unfortunately, only one such program exists in the US: in Ashland, Oregon.

The politics and energy consciousness of Europe and the United States are different.

Pimples and All

Should hydro and wind be eligible for net metering treatment? IPP thinks so, yet net metering legislation that includes wind and hydro is often blocked by utility lobbyists. California's proposed revised version contains important changes annualizing the netting period to improve the economics for the system owner.

It can be used as a model in other states and for federal implementation. IPP thanks all members of the RE community for the tremendous amount of work expended to date.

Where Do We Want To Go?

What does a whole loaf look like? Richard Perez's vision of energy farming oft mentioned in this magazine is what IPP wants to see too: any renewable energy (RE) resource deployed locally. These "farms" (of all sizes) would produce excess energy that could be sold at a fair price in a truly competitive energy market. This is the vision of a fully developed system of distributed generation based on RE. To quote Amory Lovins, this is nothing less than the "withering away of the utility."

How Do We Get There?

Net metering is a first step. It establishes grid access for small scale customer-generators. Interconnection standards need to be uniform and fair. These standards must be established by independent agencies. Inspection and compliance certification must be conducted by independent local authorities. Equipment manufacturers must build safe equipment with proper protection features built-in. System designers and installers must be qualified and understand the basics of electrical safety and trade practices.

Let's Get Weird

Imagine this twisted scenario of a hypothetical highway system. The road builders write the rules for using the road. Only large trucks belonging to approved companies are allowed to use the roadways. No cars or motorcycles are allowed. The road builders and maintenance crews also patrol and write tickets and remove "offending" vehicles from the road. The road builders collect huge amounts of money from the beneficiaries (users) of the road and spend it lobbying and advertising that this is the best and safest structure for all. The public buys it lock, stock and barrel and if challenged, says, "You can't buck the road builders."

The highway system is, in fact, not like this. Safety, access, rule enforcement, building and maintenance are all handled by independent entities for the benefit of the public. The highway system serves a common carrier function. The transportation system is open to all. Shouldn't the distribution of electric energy be just as open? IPP thinks so. The public benefits of open commerce in energy will be just as important as they are for goods, services, information and entertainment.

Buyers Wanted

Equally important is the development of a market for *green energy*. Consumers must be educated (sold) about the value of RE. James Udall's article in *Home Power* Issue #64 does a great job discussing the question of value versus cost. When people perceive value, they will pay for it. The low price for commodity or bulk power is a fiction based on hidden externalities. Part of marketing green power will be to expose those hidden externalities while demonstrating the value of RE

Nuts and Bolts of Selling Power

How would RE farmers sell their excess power? In the past, the cost of meter reading, maintaining accounts, bookkeeping and check writing have been impediments to small energy producers (to say nothing of miserable avoided cost pricing). In the near future, a natural alliance between RE farmers and green marketers will develop. It is reasonable to assume that RE farmers will want to purchase power when they need it from other renewable sources. The marketers will act as brokers. The key here is that the transaction costs must be low. The use of the Internet and electronic metering will keep those costs very low. Expect electronic meters to become cheap. Each meter will be programmed with the equivalent of a Personal Identification Number (PIN). The marketer handling the account can poll the PIN on a real time basis. Consumption and production can easily be accounted for on a real time basis. The RE farmer could shift load to favourable times while making excess power available for sale at times of peak rate. Buy low, sell high! Price signals are available in real time too, and the whole process could be handled by almost any personal computer. Renewable energy has now gone beyond net metering.

Will We Get There?

IPP thinks so. It will take both technological change (lower costs, etc.) and political change (maybe consciousness is a better term). Phasing out the burning of carbon will also happen. Carbon is much more valuable as a building material than as a fuel. Are we still arguing about climate change? Remember the Carboniferous Age? Is that the Pacific Ocean coming in your window?

Californians Strike Back

Californians Against Utility Taxes (CUT) have completed the final stages of a petition drive to place an anti-nuke bailout initiative on the next ballot. Among the goals is an attempt to reverse the US$28 billion nuclear bailout that is contained in the state's restructuring law. This "largest corporate welfare package in California history" will cost every California family US$2,000. Hopefully Californians will join the citizens of Massachusetts in their success at getting an initiative on the ballot.

Interconnection Standards

The IEEE Working Group developing P929 (Recommended Practice for Utility Interface of Photovoltaic [PV] Systems) was to meet in late April and hopefully reach a consensus. The remaining issue is what a *non-islanding inverter* must do (or not do). If that can be determined, then there is general agreement among the members that the *visible physical disconnect* presently mandated by many utilities would not be required.

Slow Learners or Just Don't Care

Pacific Gas and Electric Company (PG&E) based in San Francisco, California, would rather spend money on advertising than get their act together. In May of 1996, the *San Jose Mercury* newspaper did an extensive series of articles detailing how PG&E had cut tree trimming and other preventive maintenance budgets while simultaneously giving big bonuses to top execs. The series was prompted by a high number of power outages throughout northern California. The PUC attributed the outages to poor maintenance. During the summer of 1997, a number of fires were attributed to a lack of tree trimming. In one case, a fire near the town of Rough and Ready was determined to be due to criminal negligence on the part of PG&E. In the April 10 *San Francisco Chronicle* appears an item titled:

GUILT BY ASSOCIATION: Pacific Gas and Electric Co. will be fined for letting an affiliate use its name and logo in ads without clearly telling readers the company wasn't the familiar PG&E, the state Public Utilities Commission said yesterday. The PUC said PG&E Energy Services, an unregulated energy provider and affiliate of the utility, broke the rules by writing illegible disclaimers — in small type, running vertically up the side of an ad, and in colors that did not stand out from the rest of the ad — in several newspaper ads.

This is a blatant abuse of market power by the utility. Even if they completely changed the company name, the affiliate still represents an abuse of market power since the ownership remains with the parent company. Remember, the parent company is receiving a massive bailout from the rate taxpayers.

These phony "competitive" affiliates shouldn't fool anyone. So, watch out, they will be knocking on your door offering "green" energy and services.

Net Metering Update

Tom Starrs, a national leader and authority on net metering legislation, shares some recent good news with *Home Power* readers:

1. The Clinton Administration's restructuring plan (available on the DOE website) calls for national net metering and standardized interconnection for all renewable generating facilities 25 kilowatts or smaller. No specific bill language yet — and as we all know, the devil is in the details.

2. The National Association of Regulatory Utility Commissioners (NARUC) passed a resolution urging state commissions and legislatures to adopt measures to make net energy metering available to small scale renewable generating facilities, and requesting Congress and the FERC to identify and remove any barriers to state implementation of net energy metering.

3. There were important pro-net metering decisions in Maine (favorable rules for implementing net metering under retail competition), Iowa (Iowa Utilities Board abandons proposed rulemaking that would have abolished net metering) and New York (Public Service Commission issues broad ruling on net metering implementation, rejecting utility indemnification and insurance requirements and rejecting interconnection requirements beyond those already negotiated).

4. Governor Locke signed Washington State's net metering bill into law on Friday, April 3. This is arguably the best net metering law in the country, for the following reasons:

- It extends net metering to solar, wind and hydro (many recent laws have been solar-only).

- It applies to all customer classes and to facilities generating 25 kilowatts or less (other laws are residential-only, or have a lower 10 kilowatt limit).

- It allows month-to-month rollover credit for any excess generation (only New York has this requirement, though Maine is considering it).

- It specifies uniform interconnection requirements based on recognized national standards (only Maryland and Nevada have similar requirements, which reduce the utilities' ability to "gold plate" interconnection requirements).

There's More

IPP member Andrew Perchlik shares this good news from Vermont: "I am happy to announce that H605, Vermont's net metering bill passed! It is now waiting the Governor's autograph which is 99% guaranteed. Due to Vermont's rural farm culture our net metering bill is a bit different. The law allows businesses and residential customers to establish intertied systems of 15 kilowatts and less, but farms are allowed to go up to 100 kilowatts. This allows farms that have or will have methane generators (created by anaerobic digestion of ag waste) to take full advantage of that technology. Other highlights are: eligible systems include PV, wind, fuel cells running off a renewable fuel or a farm system generating electricity from anaerobic digestion. Excess power generated by the system is credited to next month's bill.

"Any excess left at end of year is granted to the utility. Utilities must provide intertie on first-come, first-served basis until the cumulative generating capacity of net metering systems equals 1.0% of the utility's peak demand during 1996."

And in Oregon

Oregon SEIA and others are working on net metering. Oregonians, this is your time. If you have not already gotten involved, contact the Oregon Solar Energy Industries Association.

A Blatant Plug

A just published 1997 Special Issue of *The Energy Journal* from the International Association For Energy Economics (IAEE) is devoted wholly to distributed generation. The special issue titled, "Distributed Resources: Toward a New Paradigm of the Electricity Business" contains nine articles exploring the many aspects of this topic. Setting the tone, the introduction begins, "Distributed Resources (DR) is an intriguing subject. It was certainly instrumental in creating the electricity industry at a time when transport costs were simply unbearable. Rejected later by those companies that it helped create, DR is now staging a comeback."

Jay Morse, a California PUC regulator authored one article, "Regulatory Policy Regarding Distributed Generation by Utilities: The Impact of Restructuring" which examines DR issues from a regulatory point of view. He first looks at DR within the context of the integrated utility of the past and then reexamines the issues for the restructured electric utility of today. The outcomes are quite different. Highly recommended reading.

Hit Piece

RE must be making the Carbon Cartels nervous. For a really bad time, read the March 9, 1998 article in *Forbes* magazine by Kelly Barron. The title, "I'm greener than you — Solar panels on roofs are due for an encore. Same hype, same lousy economics" jump-starts the reader either to nausea or reactionary glee. The article quotes four members of the PV industry in a context that is very unfriendly. It's hard to believe they would knowingly contribute to such a nasty article.

Guys, please tell me you were tricked! Thanks to IPP member Allen Carrozza.

There is no doubt that oil, gas and utility lobbyists are going after renewables. Check out the *Sustainable Energy Coalition Update* — it's full of important stuff. And please send publisher Ken Bossong a donation.

IPP Online Soon

IPP member Bill Lord has offered to design and host an IPP webpage. Bill, thank you very much. The site will include a statement of purpose, a list of members with access data, IPP Logos for downloading, the legal status for RE systems in various states, a list of states in net metering flux with access to organizers and a summary of PURPA, FERC and what all this means to small RE producers. The IPP web site can be a national forum and organizing focus point for small scale RE producers.

Special Offer to IPP Members

IPP members that advertise in *Home Power* and display the IPP logo will receive a 10% discount on the ad cost. Don't miss this opportunity to save some money on advertising and promote IPP.[1]

Editor's note: This article first appeared in *Home Power* Issue #65 (June/July 1998) and is reprinted with permission of the author.

Seeking Our Own Level

PAUL CUNNINGHAM

This second issue of *Home Power* magazine gives me the opportunity as Hydro Power editor to wax philosophical. A chance to put aside thinking about the *hows* of generating electrical power from water and to reflect on the whys, by still waters, of course.

Around a decade or more ago a certain realization was taking hold. Yes, we could escape the prescribed route of greater specialization, consumerism and urbanization that North American culture had mapped for us. The ultimate metaphor for carving out our new lifestyle from the social and spiritual wilderness was to generate our own electricity from wind, sun and water. Home Power. We were and are literally putting the power back into our own hands. It was a matter of the amperage and the ecstasy. Becoming more conscious of our energy generation and consumption also brought the realization that we really needed very little electricity to be comfortable.

So Where Are We Now?

This is difficult to assess since the people involved are by their situation a very decentralized group. Yet, I receive letters from all over the world from people who know something about head and flow, nuts and volts — and also from those who don't, but believe in the magic of turning water into electricity. The truth is, we are everywhere. We are part of an unnoticed, but vital and growing, network of people who are interested in generating their own power. And now this spectrum has broadened to a great degree.

Reasons for small scale power generation range from the practical (beyond the commercial power lines) to the environmental (small scale generation is less harmful than megaprojects or nukes). The original trickle of backwater hydro power enthusiasts has swelled. Water, of course, is not deterred by obstacles — it flows over them, wears them down through time and seeks its own level. Something like this is happening with the alternative energy movement in general.

The part that is successful has persevered and attracted a following on its own terms. A very interesting aspect of this movement is what can be offered to developing countries. Progress does not have to mean expensive large projects and centralization of power generation. Individual people can master this simple, small scale technology. This mastery will dramatically change their lives. Just a little energy production can produce vast improvements in the quality of life. Alternative energy can provide lights for a village to work or read by, or power pumps to move water for drinking or irrigation, or power tools for cottage industries. The possibilities of alternative energy are endless and revolutionary. The surface has barely been scratched.

So Let's Change

Clearly the world needs a new blueprint for development and change. Alternative energy is definitely part of this new blueprint. At least, there is now some groundwork in this field that proves its viability. This, alone, is an accomplishment.

This magazine will help in a technological and philosophical exchange of ideas. *Home Power* is a forum for small scale alternative energy. Right now there is no other publication that seriously addresses the requirements and interests of people involved in personal power production. We need a higher profile if we hope to be one of the keepers of the light.

It is unclear why home-sized water power, in particular, is so little known. It is true that other forms of comparable energy sources receive far more attention. The supreme reliability of photovoltaics and the romance of wind power are well established. Somehow the use of residential-sized hydro power has been largely overlooked. Part of this is likely due to the sound of the output figures. Although a water power system may produce 100 watts of power 24 hour per day, it sounds like so much less than a PV (or wind) system that has a peak output of 1,000 or 2,000 watts. Yet the water system could easily produce more total power

output over a given time span. And be much cheaper.

I read recently in a magazine (*New Shelter*) a comparison of three types of alternative energy systems. It was stated that "experts agree" that a hydro site capable of less than 500 watts continuous output is simply not worth bothering with. It is safe to say that a wind or PV system with this level of output would be at least a five figure investment. My own household operates on a maximum of 100 watts of continuous power input and runs quite successfully on less when water flow drops. Please understand that all forms of alternative energy technology are site-specific. At any given location there may be compelling factors that favor one form. This site-specific nature still doesn't explain the low proliferation of water power.

This discussion does not imply competition between the various forms of alternative energy. The situation is one of cooperation rather than competition. Many times more than one type of power generation can be used to produce a hybrid system that is both more reliable in output and more cost-effective than a single source. The point being made is simply that the very useful source of water power should not be overlooked.

So far no large business has attempted to develop the personal-sized hydro market. The advantage to the small manufacturer like myself, of course, is that we can still remain in business. The small hydro market has such a low profile that raising it by any means would probably be helpful to all involved. At present, none of the few small manufacturers has the business machinery to aggressively promote their product or to greatly increase production if it was required. The industry is in its infancy.

A Look Forward

Improvements in magnetics and electronics make possible devices that would be a quantum leap ahead of the present day offerings. Higher-frequency generation using the new super magnets, coupled with solid state switching, could create cheaper and more efficient machines. Although more advanced machines are not strictly needed, a certain amount of R and D is necessary to produce any product. This will continue and is healthy for both the industry and the consumer.

But thus far the machinery itself is not the limitation on its use. The consciousness of the market is controlling the growth of alternative energy at this time. This became very clear to me when I first started my business. Most of my sales went to the US West Coast even though my location is in Atlantic Canada. The main work needing to be done is increasing the awareness of potential alternative energy users. So you corner the market. What if there is no market? I believe the market is unlimited, but no one has noticed. This is certainly the case in developing countries. Most areas have little or no power. And these people are not likely to be reading our English language publication.

So This Is The Challenge!

To spread the word any and every possible way. This is why we are here with *Home Power*. Hopefully this will set in motion the realization that we (and our planet) will benefit more from small local power systems than the centralized, capital-intensive types.

Editor's note: This article first appeared in *Home Power* Issue #2 (January 1988) and is reprinted with permission of the author.

Clearing the Air—
Home Power Dispels
the Top RE Myths

IAN WOOFENDEN

Home Power's position in the small scale renewable energy (RE) community ensures that we hear it all, every day. Along the way, we've found that there's more than a little misinformation out there. Many RE myths are so widespread that they represent bona fide hindrances to the increasing use of these important technologies. This article is our collective debunking effort, in the interest of clearing the air.

Myth: Solar Living
Means Sacrificing Conveniences

Our solar home has all the conveniences that Karen and I want. Solar energy provides the electricity to run computers for our work; it pumps our water from the well; it entertains us with video and audio; it washes our clothes; it reheats our food and drinks in the microwave; it powers our refrigerators and freezers; it powers our ham radio, telephone and Internet communications; it runs our power tools and it lights up our nights. Solar heat cooks our food, heats our house and provides hot water for washing our clothes, dishes and bodies.

The only "convenience" we don't have is paying that monthly utility bill.

— Richard Perez
(richard.perez@homepower.com)

Myth: Wind Turbines Kill Birds

Do wind turbines kill birds? Some do. Is it significant? No. The question has been studied a great deal for utility-scale turbines. These massive turbines kill fewer than two birds per turbine per year. While no one wants to kill any birds, this number is dwarfed by the number of birds killed by habitat destruction, pollution, domestic cats, electrocution by utility lines and collisions with windows, cars and buildings.

For example, in the United States agricultural pesticides are conservatively estimated to kill 67 million birds per year. Wisconsin Department of Natural Resources research suggests that rural free-ranging domestic cats in Wisconsin kill about 39 million birds each year. The windows in your house probably kill more birds in a year than the average wind turbine.

What about home-scale turbines? No studies have been done on these turbines, and researchers do not consider the issue significant enough to study. Compare a utility-scale turbine with a home-scale turbine. Even ignoring the massive towers, a typical utility-scale turbine is 50 to 200 times larger than a typical home turbine in swept area. This in itself is enough to answer any concerns about birds and a wind turbine at your home.

Birds must navigate through a wide variety of obstacles in their flying careers. Wind turbines pose no special hazards to them, and are in fact easier to notice and avoid because they move. In my 20 years of living with wind turbines, I've seen birds regularly alter their courses to avoid our turbines. Birds sometimes even perch on our turbines' stopped blades, but they leave as soon as the wind comes up and the blades start rotating.

Everything humans do has an impact on other people and on the environment. If you're looking for an energy source with no impact, good luck. Obviously, wind farms need to be sited intelligently, not directly in major bird migration flyways. But before we stop installing wind turbines because of a few bird kills, we should get rid of cars, buildings, utility lines and cats…

For more information on wind turbines and birds, see awea.org/faq/sagrillo/swbirds .html.

— Ian Woofenden
ian.woofenden@homepower.com

Myth: Solar Panels Make Electricity from the Sun's Heat

There are two major types of solar panel technologies. When it comes to how they work, they couldn't be more different from each other.

Solar hot water panels, also known as solar thermal panels or solar *collectors*, capture the sun's heat to provide hot water for domestic use or home heating. These are large, dark, rectangular panels usually measuring around 4x8 feet. They look like very shallow rectangular boxes and have been around and in use on residential rooftops for decades.

The second type of solar panel is the photovoltaic (PV) panel, also known as a solar-electric panel or module. These smaller and much lighter-weight panels use the sun's light to make electricity via what's known as the *photovoltaic effect*. PV modules perform best in cool temperatures under bright sunlight. They come in all different sizes (including some that are cleverly disguised as roofing materials) and are turning up in a wide variety of residential, commercial, industrial and scientific applications.

So you can get hot water from the sun's heat and electricity from the sun's light. If you've got sunshine, there's nothing keeping you from choosing both!

— Scott Russell
scott.russell@homepower.com

Myth: It takes More Energy to Build PVs than They Can Ever Produce

Some skeptics of solar energy claim that it takes more energy to make a photovoltaic module (PV) than it can ever produce in its lifetime. The truth is that PVs typically recoup their embodied energy in two to four years. According to an article published by the National Renewable Energy Laboratory (NREL), today's single and multicrystalline modules have an energy payback of about four years, and thin-film modules about two years.

Most PV modules in the field are made from hyper-pure crystalline silicon. Purifying and crystallizing the silicon consumes the most energy in making these PVs. Thin-film PVs are made from considerably less semiconductor material, and therefore have less embodied energy in them. Most of the energy consumed is in the thin-film surface. The aluminum frame on any PV accounts for about six months of its payback time.

Solar energy is an amazing technology considering that PVs go on to produce clean, pollution-free energy for at least 25 to 30 years after they have achieved payback. For more information on energy payback, see the National Renewable Energy Laboratory's website (nrel.gov) and Karl Knapp and Theresa Jester's article titled "PV Payback" in *Home Power* Issue #80.

— Eric Grisen
eric.grisen@homepower.com

Myth: Burning Wood as Fuel is Bad

Plenty of bad things can happen when burning any carbon-based material. But wood is renewable in the short term, which makes it one of the best carbon-based fuels for heating. CO_2 is a problem with burning nearly anything. In the case of wood, the same amount of carbon is released by burning as would be released by the natural decay of a fallen tree — there is no net increase in atmospheric carbon. With fossil fuels, the common alternative to wood fuel, the carbon is permanently locked up in the fuel unless burning lets it out, causing an increase in atmospheric CO_2, a proven cause of global warming.

There are negative effects of burning wood, mostly from particulates that get released. But using an EPA-certified wood heater will minimize this problem. There is always some kind of negative impact from creating heat. The goal of the considerate and responsible energy user should be to minimize these impacts, helping our world to become as sustainable as possible. The best way to heat is with the sun. But if you have to burn something, either make sure it is renewable or that it is made with a renewable resource, and be sure it is done as efficiently as possible. See John Gulland's article on efficient and clean use of wood as a fuel in *Home Power* Issue #99.

— Michael Welch
michael.welch@homepower.com

Myth: Solar-Electric Module Production is Toxic to the Environment

A while back, there was a media barrage claiming that photovoltaic (PV) manufacturing was extremely hazardous to the environment. PV manufacturing does require the use of chemicals that are designated as toxic by the US Environmental Protection Agency (EPA). Employee safety is paramount during the manufacturing process, and chemicals used must be disposed of in an environmentally sound manner.

The federally funded National Renewable Energy Laboratory (NREL) researched the media claims and concluded, "By using well-designed industrial processes and careful monitoring, PV manufacturers have minimized risks to where they are far less than those in most major industries. All of these risks fall well within the range already protected by OSHA and similar regulations."

A thorough analysis of the environmental impact that various energy sources have on the environment must take into account the net effect of a given source over the source's operational lifetime. When you compare the environmental impact of PV technology to traditional energy sources like coal and nuclear energy, PV comes out on top, hands down.

Nukes produce nuclear waste, and even after spending billions of taxpayer and ratepayer dollars, no acceptable disposal solution has been brought to the table. Fossil-fuel based energy sources like coal produce air pollution over the power plant's entire operational lifetime — as long as it's running, it's polluting. Burning coal releases sulfur dioxide, which results in acid rain; nitrogen oxide, which results in smog; carbon dioxide, which results in global warming; particulates, which result in lung damage and an array of heavy metals like arsenic, lead and mercury, which result in birth defects and brain damage.

On the other end of the spectrum, PVs produce no emissions and require no use of finite fuel sources. PVs manufactured today are expected to be producing energy 50 years from now. PVs offset all the energy used to manufacture them (embodied energy) in two to four years in most locations. Fossil, nuclear, or solar — which energy source would you want in your backyard?

— Joe Schwartz
joe.schwartz@homepower.com

Myth: Microhydro is Bad for River Life

The impact of microhydro on fish and other river life is tainted by association with blatantly destructive, large-scale hydro, which seriously impedes fish movement, changes stream temperatures and flow rates, slices and dices aquatic life and even drowns entire ecosystems.

Microhydro does none of these things — if appropriate precautions are taken. There is always going to be some negative impact, but that can be said for nearly every human activity — even walking down a forest trail. Some misguided folks do not consider the impacts of what they do, and they give a bad reputation to those of us doing similar things in a more caring and respectful manner.

The idea is to minimize the impact of microhydro by following some simple rules:

- Always leave enough flow in the stream bed for aquatic life.
- If migratory fish use your stream, make sure that they and their fry can swim past your diversion and cannot be drawn into the penstock intake.
- Always put the diverted water back into the same stream bed in a way that does not cause erosion.

— Michael Welch
michael.welch@homepower.com

Myth: Solar Electricity is Too Expensive

There is a huge public misconception that solar energy is simply too expensive to bother with. The reality is that, both on and off-grid, solar energy is cost-effective in many applications.

Right out of the gate, it's important to understand that on-grid, a substantial amount of "smoke and mirrors" is going on behind the scenes, making true energy cost comparisons unfair at best. The historical trend shows US federal energy subsidies favoring mature energy sources like coal and nuclear over renewable sources by a factor of 100 to 1. A report based on US Department of Energy (DOE) data by the Congressional Research Service (CRS) states, "Because the great bulk of incentives support mature fossil and nuclear equipment, the existing subsidy structure markedly distorts the marketplace for energy in a direction away from renewables."

The bottom line is that renewable energy appears to be more expensive than traditional electricity generation sources, but the reality is that you pay the difference every year come tax time. If you include the costs of increased pollution, habitat destruction, health care costs etc., then RE looks even better. Fortunately, many individual states are doing what the feds refuse to do, and are implementing rebate programs for renewables that serve to even out the financial playing field a bit. For some great economic analyses of the cost-effectiveness of grid-tied PV, see the article by Greg Bundros in *Home Power* Issue #99 and the article by Paul Symanski in Issue #100.

Off-grid, people have been realizing the financial advantages of solar energy for more than a decade. Property beyond the reach of the utility grid is typically undervalued and a great investment. We're not necessarily talking about living "out in the sticks." A good rule of thumb is that a solar-electric system costs less than a utility line extension of a quarter mile or more.

I had the local utility provide me with an estimate for running a line to my off-grid homesite (though I was never going to take them up on it!). They came up with a cost figure of US$32,000. I used this estimate as leverage when I purchased the property, which substantially lowered the seller's asking price. From day one, renewable energy technology saved me over US$10,000 compared to bringing the grid in. How's that for an incentive!

— Joe Schwartz
joe.schwartz@homepower.com

Myth: You Can't Use Solar Energy in Far Northern Latitudes

Solar energy can and does work in northern latitudes. A trip to any well-designed passive solar building can be one of the most uplifting experiences in the cold winter months because of the warm, cozy atmosphere it affords. Every square foot of south-facing insulated glass can let in the heat equivalent of about a half gallon of heating oil from the sun each heating season. Cover the glass with insulating shades or shutters at night, and the heat equivalent can increase to nearly a gallon for each square foot of window.

There are too many examples of the successful use of solar energy in northern latitudes to be included here, but hundreds of solar home owners in far northern latitudes have opened their doors in the American Solar Energy Society's National Solar Tour (ases.org). *Home Power* magazine has been bringing you articles about successful solar-electric systems in Canada and the northern United States for the past seventeen years. Germany, the world's second largest user of electricity generated by PV modules, is not located in the sunbelt, but rather at 48 to 54° N latitude.

Obviously, the largest obstacle to using solar energy in the north is the short, cloudy days of winter. Annual net metering of PV systems has really helped overcome this obstacle for on-grid solar-electric systems by providing a year's energy *storage* (in terms of dollars and cents from a billing perspective). The long, sunny days of summer can directly compensate for the shorter days of winter in northern latitudes.

Something interesting to think about is that the peak electrical loads in many northern cities (such as my home of Burlington, Vermont) have shifted from the winter months to the summer months over the past ten years. This shows that there is ever increasing potential of solar electricity in northern latitudes to complement the passive solar and solar thermal systems that have been working for the past 20 or more years up north.

— AJ Rossman
aj@drakersolar.com

Myth: Lead-acid Batteries Wind Up as Toxic Disasters in our Landfills

Hardly any other industry does a better job at recycling than the lead-acid battery industry, and this includes aluminum, glass, paper and plastics. More than 90% of spent battery lead is recycled, which is two to four times higher than many other major recyclable commodities. And 60% of the lead used in manufacturing lead-acid batteries is derived from recycled lead. Most of the lead used in your car's battery has probably ridden around in three or four other cars before it got to yours.

Worn out lead-acid batteries are accepted for recycling by all outlets that sell these batteries — it's the law. From there the batteries are broken open, and the lead is removed and resmelted for reuse in new batteries. The only way a lead-acid battery winds up in a landfill is if a careless user dumps it there. So don't break

the recycling chain — return your spent batteries to a dealer for recycling!

— Richard Perez
richard.perez@homepower.com

Myth: Grid-intertied PV is Hazardous to Utility Line Workers

Although this may be one of the most pervasive myths in the electricity industry, I was unable to locate a single documented instance of injury or death to a utility worker from a grid-intertied inverter. The reasons for this are twofold — modern inverter design and line worker safety protocol.

Inverters are perhaps the most highly scrutinized piece of electronics used in residential applications. Their safety and proper functioning are certified by some of the same agencies that verify the safe operation of all the other appliances in your home.

Inverters for use in grid-intertied systems are required (by IEEE, the NEC and UL) to disconnect from the grid for any number of conditions. These include grid outage, high or low voltage, high or low frequency and inverter malfunction.

Inverters are required to have several redundant safety devices built into their electronics to ensure that they disconnect from the grid if anything at all is wrong. Nonetheless, utility companies and line workers are quite safety conscious and leave nothing to chance.

Line workers are trained to always ground any potentially energized conductors when performing utility line maintenance. In addition, grid-intertied systems are routinely required to have a safety disconnect available for the line worker's use to lock out any solar electricity generation from being backfed onto the grid.

Line worker safety protocols make a great deal of sense. During utility outages, many people use engine/generators to keep the electricity on in their homes and businesses. Most engine/generators do not have the intricate electronics that inverters have to ensure line worker safety. If they are not correctly hooked up with a transfer switch to isolate selected circuits in the home from the utility grid, the generator can backfeed electricity to the grid through the utility's transformer, which converts it to extremely high voltage.

Line workers have been killed by engine/generators, so it's a good thing they practice safety rigorously. In fact, the problems with engine/generators are the reason utilities have been so cautious about allowing any other customer-owned generating sources on their lines at all.

Since inverters have such a strong safety record, some day soon they will be a common and accepted part of many home electrical systems. They will outlast the urban myths of line worker lore. For a more thorough discussion of utility-intertie inverters and how they work, see *Home Power* Issue #71, page 58.

— Linda Pinkham
linda.pinkham@homepower.com

Myth: All Solar Heating Systems Need a Backup Fossil Fuel Energy Source

While it is true that most solar heating systems have a conventional backup heating system, it isn't absolutely necessary. Fossil fuel heat as a supplemental system is a cost, financing and comfort decision. Many solar energy heating systems rely on the renewable resource of wood for any heat not supplied directly by the sun.

A combination of passive and active solar energy collection is probably the easiest and most cost-effective way to avoid a conventional backup system. A superinsulated passive home design in a sun-friendly climate can provide all but a small fraction of the energy needed to heat a home. An active solar heating system typically stores heat in a large storage tank (many people use an indoor pool) for the times that the passive system is unable to collect enough energy, or a severe storm calls for more heat than normal. A PV system provides the required electricity. This type of design is not the norm by far — it's just a little too expensive upfront for most people — and it might require the owners to put on a sweater indoors a few times a year.

The expense of going 100% solar and the possibility that the home might fall to 60°F or so in rare circumstances are the reasons that most solar homes have a conventional backup. Another factor that looms large for many people is that mortgage bankers are very nervous about lending money on homes that fall out of the conformity they are familiar with.

— Chuck Marken
chuck@aaasolar.com

Myth: Hydrogen Fuel Cells are a Renewable Energy Source

Hydrogen fuel cells produce DC electricity from hydrogen. They do this cleanly and quietly. But where does the hydrogen come from? Though hydrogen is the most common element on earth, unlike sun, wind and falling water, it is not freely available. It must be stripped out of hydrocarbons or split out of water. These operations take energy, and the actual energy source may not be renewable at all.

Hydrogen can be thought of as an *energy carrier*. We use some energy to get it out of hydrocarbons or water, and then we get the energy back when we run the hydrogen through a fuel cell or engine. Every conversion of energy has an efficiency cost and an equipment and maintenance cost. If hydrogen fuel cells have a place in renewable energy systems, they must be a step forward in terms of cost, efficiency and environmental friendliness. The jury is still out on this issue.

In renewable electrical systems, hydrogen fuel cells might replace two different components that we use today — generators and batteries. Many people use gasoline, diesel or propane-fired generators as charging or backup sources in off-grid or on-grid RE systems. Fuel cells could be a quieter, cleaner answer, even if they use nonrenewable fuels.

To replace batteries in RE systems, you need two other components besides the fuel cell. First, an electrolyzer is needed to split hydrogen out of water, using your surplus renewable energy. Then you need a hydrogen storage system — not a simple proposition.

Any new technology takes time and money to develop. Hydrogen fuel cells may play a role in RE systems in the future. But the energy sources that power them should be the sun, wind, falling water and the like. Otherwise we are just pinning our hopes on more nonrenewable energy, with a high-tech twist.

— Ian Woofenden
ian.woofenden@homepower.com

Editor's note: This article first appeared in *Home Power* Issue #100 (April/May 2004) and is reprinted with permission of the author.

Stimulating the Picohydropower
Market for Low Income Households
in Ecuador — Executive Summary

ENERGY SECTOR MANAGEMENT ASSISTANCE PROGRAMME (ESMAP)

The main aim of this World Bank-ESMAP project has been to pave the way for picohydro to become accessible to low-income households in Ecuador, with the view to replicating the process in the Andean region and other developing countries.

This has been done through five main activities:

1. Assessing the main experiences in picohydro technology and market developments, looking particularly at Vietnam and the Philippines;
2. Reviewing the existing use of microhydro in five Andean countries and assessing their potential picohydro market;
3. Establishing 31 pilot picohydro projects in two provinces of Ecuador with full participation of the end users in local communities;
4. Encouraging a sustainable local infrastructure for supporting picohydro technology in Ecuador through training local technicians and bringing local dealers forward to consider picohydro as a business venture; and
5. Conducting a rapid rural appraisal (RRA) in the communities before and after the installation of the picohydro pilots to consider the technological and social impacts in the Ecuadoran situation.

Main Outputs

There are five main outputs of this project.

First, the project builds on the findings from an associated UK Department for International Development (DFID) project, which evaluated the successes and lessons to be learned from the Vietnamese picohydro experience and assessed the factors that have led to the emergence of a picohydro market in the Philippines following the import of equipment from Vietnam.

As part of this process, an investigation was done of the options for technology transfer of picohydro to Ecuador and how improvements in equipment quality can help to improve the sustainability of the market.

Second, a rapid review of the extent of existing use of microhydro and picohydro in the Andean countries (Ecuador, Peru, Bolivia, Colombia and Venezuela) was made, including assessing the type of market, installed capacity

Market Size for Picohydro in the Andes

Country	Non-Electrified Rural Households	Technical Achievable Number Of Households That Could Use Picohydro	Range Of Genuine Household Market Based On Capacity And Willingness To Pay
Bolivia	515,815	355,000	55,000–109,000
Peru	462,783	671,000	98,000–197,000
Ecuador	249,199	137,000	16,000–32,000
Colombia	127,343	39,000	7,000–14,000
Venezuela	72,170	28,000	4,500–9,000
Total	2,427,310	1,230,000	180,500–361,000

ESMAP

and costs of the technology. An assessment of the potential market and the institutional barriers to the diffusion of picohydro in these Andean countries was also made. Based on rural electrification and income data as well as estimation of hydrological resources in the various regions of the five Andean countries, the estimates that resulted from this assessment are summarized in Figure 1.

Third, 31 picohydro pilot projects were established in five villages of two provinces with varied topography and natural environment. Ten pilots were installed in Chimborazo in the Andes Mountains and 21 in Napo, in the lowland Amazonian rain forest. These projects have provided lighting and reliable AC power for a total of 193 people; and in the monitored first six months of operation, although there were two instances of generators burning out (these were replaced under warranty), the units are all operating and continue to be used even in the village that has subsequently had a grid connection.

Fourth, to build the foundation for a sustainable local commercial infrastructure for picohydro technology in Ecuador, a training course was given for installation and maintenance aimed at engineers and technicians. In addition, discussions about business opportunities for picohydro were conducted with potential importers and companies that have the rural-based commercial structure in place to eventually act as dealers. This resulted in high levels of interest from most in the private sector, with the view that even without any marketing about 20 picohydro systems could immediately be sold in one year by each dealership for a very small initial investment. However, there was a requisite that more technical information be forthcoming and support for importation be put in place, because it was recognized that if the units were made in Ecuador, the price would be higher than if they were imported from Vietnam. Also, one of the larger electrification companies thought that despite picohydro being an economic service for isolated households, the company could not make it into a profitable business.

Fifth, an activity on rapid rural appraisal (RRA) of the pilot projects ran throughout the project, which established a baseline for the communities chosen, monitored the picohydro equipment for more than six months of operation and conducted a social impact assessment at the end of the project.

Experience with Picohydro Technology

Typically, picohydro systems have a capacity of between 200 and 1,000 watts of electrical power, but the term *picohydro* can include systems up to five kilowatts, with a range of turbines for varying heads (low head propellers, medium head Turgo or crossflow runners and high head Pelton wheels). Generally, they are used for domestic electricity applications such as lighting, television and radio and battery charging. The units are small and cheap and typically are owned, installed and used by a single family, hence, their commonly used name of *family-hydro*.

In Vietnam, China and Nepal picohydro technology has become widely available at an affordable cost to rural households, with turbine prices in the range of US$25 for 200 watts to US$1,000 for three kilowatt community Pelton systems. The technology that has fueled this market development has been mainly a mix of locally-made low and medium head units, although the experience has been that units from China are inefficient, unreliable and sometimes unsafe and have a life of less than two years. Generally, the market mechanism has been a "cash and carry" model in which technical information is spread by word of mouth.

There are some other countries that have experienced picohydro market development, primarily in Southeast Asia (e.g., the Philippines), India and South America. There has been only minor uptake of the low head technology in Africa to date, but there remains a massive unexploited potential throughout the developing world and in niche markets in North America and Europe.

Vietnam has seen the purchase of most picohydro schemes outside China, and it is estimated that approximately 120,000 units have been installed since the late 1980s, most in the northwest corner near the border with China. The Philippines has recently experienced a small, but growing activity in the development of low head picohydro turbines for 220 volts through the import of equipment from Vietnam. The distance of the Philippines from mainland China has meant that the poorer-quality equipment that has flooded the Vietnamese market has not been able to penetrate the Philippines, so there has been a more controlled development of family-hydro in that country.

On the basis of a project undertaken for DFID between 2002 and 2004 in Vietnam and the Philippines called the "CDM Pilot Project to Stimulate the Market for Family-Hydro for Low-Income Households," a study of the two different market mechanisms that have taken place in these countries has been gained for this ESMAP project.

The following are the main concluding points and lessons learned from this project that are important in the Ecuador situation:

Compared with other small scale renewable and conventional energy options for rural areas of developing countries, picohydro is one of the most immediately affordable sources of electricity (even for the poor).

Vietnam has seen the most development of picohydro for small amounts of power for family use, with about 120,000 units deployed, but not all of this development has been sustainable because of poor-quality products being employed by rural people with annual incomes typically of only US$300.

Nevertheless, the Vietnam market for picohydro remains strong, and even considering pressures that tend to reduce the market (grid extension efforts and promises made), it is still expected to be in a range of 20,000 to 25,000 units per year.

The Philippines had virtually no low head picohydro development until units began to be exported from Vietnam in the late 1990s. Most of these units are of higher quality and have been installed with competent engineering support and the use of proper civil works, using funding support from local (and national) government and nongovernmental organizations (NGOs). Also, to maximize the units' potential, the larger picohydro units have often been chosen to supply electricity into microgrids for clusters of households or small villages.

The Philippine market for off-grid electrification remains large (2.7 million households), and picohydro has an immediate potential market of more than 24,000 units and a total market as large as 120,000 units.

The Vietnamese and Philippine picohydro experience having been researched, an approach to analyzing market characteristics and estimating market potentials has been developed and applied directly to the Andean region.

Vietnamese picohydro technology in association with local dealers in Ecuador (also through support from the Ecuadoran government to further stimulate the local market) having been successfully demonstrated, work can now be focused on scaling up the deployment of many thousands of picohydro units in the country.

Economic Analysis of Picohydro

Picohydro technologies are part of a menu of options for bringing modern energy services to households in unelectrified areas of developing countries where the hydro resource exists. A comparison of capital and operating costs for calculating the life-cycle costs of various renewable energy and diesel genset systems was conducted.

It was found that individual picohydro is one of the most affordable sources of electricity, especially for community-based projects, with life-cycle costs for each household between US$74 and US$150 per year, compared with the life-cycle costs of solar, hybrid, wind or fossil fuel–based options, which start at US$140 per household per year. In addition, a picohydro unit can provide power at 220 volts AC instead of 12 volts (from photovoltaic–solar home systems [PV-SHS]), which then requires expensive inverters to upgrade to AC power.

The economic analysis (on a life-cycle cost basis) for good quality picohydro units shows that the technology compares favorably with other small scale (renewable or diesel-based) options for bringing modern energy services to households in unelectrified areas of developing countries. As far as being affordable, picohydro can in many circumstances be within the reach of the poor (using the definition that the poor live on less than US$2 a day) and is valuable because AC power that can be linked to income-generation activities is produced.

Opportunities for Picohydro in the Andean Region

Analysis has shown that Ecuador has a good potential for exploiting picohydro power resources and a history of micro- and minihydro on which to base the new development of picohydro. The potential actual household market for picohydro is estimated to be a minimum of 16,000 in Ecuador out of a technically achievable total of 137,000 unelectrified residences that are off-grid and near the required hydraulic resources.

The neighboring Andean countries show even more potential, especially Peru and Bolivia, with a minimum of 98,000 and 55,000 households respectively, estimated to be the market in those countries. From the DFID study that cofinanced this Energy Sector Management Assistance Program (ESMAP) project and analyzed the global market demand for picohydro systems, the genuine household market potential for low head picohydro was found to be 285,000 in the five countries of the Andean community, which is similar to the range of 180,000 to 360,000 estimated in this project, which could represent between 7% and 15% of the households yet to be electrified in these countries. This compares with a global potential of about four million units and 740,000 estimated for all of Latin America

(Brazil has the highest potential because of its relatively larger population).

Experience in the picohydro sector in Asia, in which hundreds of thousands of systems have been sold in the past 15 years on a cash and carry basis, has shown that it is also realistic to deploy that many picohydro units in Ecuador and the Andean region. However, a major lesson learned is that because of the poor quality and limited life span of cheap equipment, less than half of the 120,000 units installed in rural areas of Vietnam are actually still operating.

Recent developments in the Philippines and now through this project in Ecuador have shown that there may be a model for the sustainable deployment of picohydro, but that it requires higher quality equipment and engineering support. With completion of the pilot demonstrations, the costs associated with deployment (which the project supported) were US$475 per unit (including equipment, civil works and electrical cabling costs), with estimated operational and maintenance costs of US$5 per year. Given a conservative equipment life of only five years, an end user would need to save US$40 per year to purchase another picohydro system worth US$200. Given these costs, the opportunity clearly now exists for further expansion of the market for good quality picohydro products in Ecuador and the neighboring Andean region countries and other parts of Latin America.

Impact of the Project

By successfully demonstrating good quality picohydro technology through the pilot projects (with training and awareness-building activities for all stakeholders involved) and showing that a dealer-based, private sector-led approach to picohydro projects can be financially viable, a deeper understanding about how to build the local infrastructure for the commercialization of picohydro in Ecuador has been gained. To stimulate a sustainable market for picohydro, it is important that these elements be in place.

The impact of the pilots on end user beneficiaries has been carefully considered in the project. Benefits include an increase in the quality of life because of better lighting systems enabling community activities during the evenings.

Beneficiaries have pointed out that there were fuel savings and an increase in the productive output as well as increased opportunities for educational and social activities.

Through the RRA undertaken and the training of users, along with the actual implementation of the pilots, the end users overcame their initial apprehension about the technology and now have confidence in maintaining the picohydro systems and have also acquired additional skills.

It is clear that for reliable and long-lasting installation, operation and maintenance of the picohydro systems, proper site assessment (e.g., river flows during different seasons), training of both owner and operator and provision of safety and best practice guidelines are essential.

It has been found that the demonstration of picohydro technology has created a market development effect in which requests for more systems have been made in the target villages and by neighboring communities. The private sector, government and international donors can all play a part in supporting the scale-up

of picohydro deployment in Ecuador, providing an example to other Andean countries and other developing countries alike.

Recommendations for Future Actions

It has been shown that picohydro can provide services to low income families in rural areas in Ecuador. There is a segment of the rural population for which picohydro is certainly affordable, but much more awareness is required of the potential benefits of the technology over other forms of off-grid electrification (e.g. diesel gensets). However, financing support may still be required to make it possible for the poorest customers to afford a picohydro system, particularly if good quality technology is to be used and installed properly.

It is clear from the experience gained in this project that a dealer-based energy service model, with post sales maintenance support fed by currently available good quality equipment, can deliver the sustainable deployment of picohydro systems for real benefits to rural households.

Some obstacles do remain before potential dealers will entertain picohydro commercially in Ecuador, but they can be overcome. For example, a more thorough understanding of the latest technology through input from manufacturers and suppliers from Asia will give local dealers confidence, and it will be important for the conditions surrounding the importation of large numbers of picohydro units to be clarified.

Many more potential dealers would come forward if a proper evaluation was done of the actual market locations and how picohydro can add value to products and give possibilities for productive uses in these locations. At

Costs of Picohydro to Reach Ecuador Per Unit	
Equipment cost (ex-works)	$145.00
Transport (Vietnam-Ecuador)	$12.57
Customs cost per unit	$17.48
Administration of customs	$21.22
Total	**$196.27**
Note: All costs in US dollars	

ESMAP

the same time, to motivate participation in projects this study should be complemented by training and information dissemination about picohydro to potential beneficiaries.

In addition to these aspects being in place, to make this model commercially viable and to scale up the use of picohydro in developing countries, what is now required is concentration on establishing standards and certification/licensing for the products and providing technical support for feasibility studies and site-level installation, operation, maintenance and warranty. Close community liaison is crucial in determining how picohydro technology is best organized at the end user level. For example, beneficiaries may need to provide their labor and local materials to reduce initial capital costs, pay for and carry out proper operation and maintenance and understand the importance of using the technology within its capability.

Subsidy for the capital costs of picohydro technology is not a priority area except perhaps for the poorest of the poor, whereby some of the capital costs for good quality systems (US$475 per 200 watt unit in this project) would be provided by national renewable energy programs and households would be expected to meet the operation and maintenance costs and to save US$45 per year for the next new turbine.

Support is required, however, for quality assurance/licensing of the equipment from national energy ministries together with universities that have appropriate testing facilities, support is needed to stimulate the establishment of easier importation of technology and new sales infrastructure through regional bodies and the appropriate government departments and seed money is needed to help institutions set up engineering support services from multi- and bilateral agencies. That can also be done in conjunction with the private sector (dealers, rural energy service companies [RESCOs], rural banks, entrepreneurs etc.) as well as NGOs.

Editor's note: This article first appeared as ESMAP TECHNICAL PAPER #090.[1]

Total Cost Analysis for Picohydro Pilots in Ecuador

Province	Napo					Chimborazo				
Town	Tena			Cosanga		Riobamba				
Village	Zapallo	Alto Tena		Las Caucheras		Guaslan		Alto Ozogoche		All/Average
Type of Picohydro system	LH	LH	HH	LH	HH	LH	HH	LH	HH	
Number of Picohydro systems	8	4	1	3	5	1	1	7	1	31
Total Picohydro systems	8	5		8		2		8		
Per Turbine Costs in-country	$196.27	$196.27	$196.27	$196.27	$196.27	$196.27	$196.27	$196.27	$196.27	$196.27
Local transportation	$3.46			$6.25		$18.30				
Locally made flume (LH only)	$70.00	$70.00		$70.00		$70.00		$70.00		
Per Turbine Costs on-site	$269.73	$269.73	$199.73	$272.52	$202.52	$284.57	$214.57	$284.57	$214.57	$249.74
Total civil works (pipes, cement etc.)	$783.49			$651.37		$662.13				
Civil works per unit	$60.27	$60.27	$60.27	$81.42	$81.42	$66.21	$66.21	$66.21	$66.21	$67.79
Total electrical works (cables etc.)	$1,200.30			$593.77		$1,456.90				
Electrical works per unit	$92.33	$92.33	$92.33	$74.22	$74.22	$145.69	$145.69	$145.69	$145.69	$107.81
Installation costs (developers)	$700.00			$400.00		$475.00				
Installation costs per unit	$53.85	$53.85	$53.85	$50.00	$50.00	$47.50	$47.50	$47.50	$47.50	$50.50
Total Installed Costs per unit	$476.18	$476.18	$406.18	$478.16	$408.16	$543.97	$473.97	$543.97	$473.97	$475.85

Note: All costs in US dollars; LH = low head (5 feet); HH = high head (20–35 feet); The price of LH and HH units is the same

ESMAP

The Story of PowerPal Microhydro

DAVID SEYMOUR

This story begins in the middle 1990s in the northern half of Vietnam where a group of geologists employed by Palmer Resources Ltd., a small Vancouver-based company, was involved in regional mineral exploration activities. The field activities included the collection of stream sediment samples from local drainage systems.

During the course of sample collection along the drainage systems, the exploration teams often came across Chinese-made microhydro generators equipped with propeller turbines attached to primitive alternators. Sometimes they were just single units in small streams and serving one nearby family, while in other cases there were as many as 25–30 units installed along a temporary weir built across a major river during the dry season. A haphazard network of bare wires of various types and sizes then connected the alternators to the turbine owners' houses situated above the steep riverbanks.

In mid-1998, a major gold sampling fraud involving another Canadian mineral exploration company in Indonesia set off a chain of events that led to the decision by Palmer to terminate all exploration activities in South-east Asia and to lay off all the geological staff. Two of the retrenched geologists decided to investigate other business opportunities in Vietnam, and interest soon turned to the small microhydro generators that they had observed at work in that country. Nothing even remotely similar had been seen in Indonesia, the Philippines or Myanmar (Burma) where other company geologists had been exploring.

It soon came to their attention that one government research institute in Hanoi had been working for some time on a much improved alternator-equipped propeller turbine. Improvements included better castings, better quality bearings, permanent magnets for the rotors and generally a more attractive appearance. Also a simple manually-adjusted electronic load controller was added to the package. An experienced maker of fibreglass products was located for the required intake canals and draft tubes.

So the decision was made to form a Canadian corporation to be called Asian Phoenix Resources Ltd. (APR) and to raise a certain amount of working capital for market research aimed specifically at the microhydro sector worldwide. The name of PowerPal was chosen, for it was meant to imply one of the three main objectives behind the microhydro project, namely, to bring some electricity from a renewable power source (running water) to those remotely located residents of less developed countries who would never otherwise have an electric light, fan, radio or even a small TV in their homes. The other main objectives were to have something interesting and worthwhile to do during the slump in exploration activity that followed the Indonesian fraud, and to make some profits for the shareholders of APR Limited. It is fair to say that the first two objectives have been achieved, while the last one remains a "work in progress".

The main marketing tool for the company has been its website (powerpal.com), followed closely by word of mouth among its growing list of client countries, many of which now have local distributors. But the reason that the profits have been slow in coming relates to the simple fact that the people who stand to benefit most from access to even a few 100 watts of AC power are usually among the lowest income groups. Fortunately, PowerPal has come to the attention of such agencies as the World Bank (Ecuador, Laos), the UNDP (Kyrgyzstan), various local government agencies (Philippines), missionary groups (Malaysian Borneo, Indonesia, Congo, Papua New Guinea) and the like, and these agencies assist in the deployment of PowerPal units to the remote locations where they are so greatly appreciated.

In addition, APR has been able to introduce other models to the line of products that are manufactured on its behalf in Vietnam. These include the alternator-equipped high head Turgo models that have been well received in more mountainous regions; for example, some 25 units have been installed for the workers at a coffee plantation on the slopes of Mt. Kilimanjaro in Tanzania. An automatic electronic controller now has replaced the manually adjusted ELC. And interest is mounting for larger capacity models in Europe where the cost of domestic electricity has increased so much.

So the PowerPal story, born as it was out of a sudden change of circumstance for the two main individuals behind the decision to enter the microhydro sector in 1998, has made a positive impact on the lives of people and their environment in about 65 countries, from the Faroe Islands and Iceland in the north, to Chile and New Zealand in the south. The future does look bright, and APR hopes to participate in its own small way in the growth of microhydro power usage, especially in less developed countries.

Market Solutions for Environmental Problems — The Streamworks Project

SCOTT L. DAVIS

1 — Market Solutions?

Market solutions for environmental problems may sound like a contradiction in terms. However, it becomes clearer as time passes that many of the interlocking crises of modern times stem from the practice of externalizing, or *not paying*, the social and environmental costs of our lifestyles. Undervaluing energy is a mistake, not a bargain. Real costs include environmental and social costs which are too easily ignored, at least for the time being. These costs, however, do have a way of making themselves felt. Controlling waste without charging the full price of disposal, for example, is doomed to failure. When energy is undervalued, it provides incentives for certain kinds of industry that use a lot of power, at the cost of massive market distortion. However, market distortions cause market disruptions.

Pricing is an effective means of distributing resources: when the pricing mechanism is distorted, then undesirable consequences such as pollution, waste and peculiar attitudes are rife. This realization is nothing new. The Fair Trade movement makes this realization their goal. Hillary Clinton denounces free trade without a recognition of poor labour

and environmental standards. So the idea is out there. I am kind of surprised to have to point it out. However, energy does offer an excellent example of externalizing environmental costs. The examples in this book demonstrate that while clean energy is not as cheap as grid power, it is certainly affordable with the proper incentives, even without waiting for a "levelling of the playing field."

This fact, that ecological and economic justice is affordable, is a key message *Serious Microhydro* has demonstrated over and over again. It is interesting indeed to think that revolutionary changing of lifestyle and institutions may not be necessary to live within our means ecologically, so much as simply paying the true costs for our behavior. Here in this collection we have found example after example of affluent clients providing a high standard of living for themselves from local sources of power that are entirely green. Free green energy indefinitely seems to cost these people about the same as a sports utility vehicle.

At the other end of the affordability scale, you don't need to be that affluent to benefit from small water power. Indeed, since micro-hydro power (where available) is the most cost-effective renewable energy technology, there are many examples showing microhydro providing reliable and low cost power to people in remote areas.

2 — Barriers

Much microhydro power potential remains to be appreciated and developed. One may wonder why, if it's such a great idea, more hasn't been done already. There are some barriers to microhydro development that are specific to microhydro technology. For one thing, microhydro is for some reason the shyest renewable technology. Where solar and wind have done extensive marketing, true knowledge about small water power is even still remarkably hard to come by. There has never been a market survey of microhydro, as mentioned before. Myths get propagated and go unanswered, leaving a confused public.

Microhydro also shares barriers with other renewable technologies. At a larger scale, energy is a confused topic. This topic is "confused" by outright propaganda from energy companies and government, abetted by the market distortions that people have come to expect in order to keep energy cheap enough to waste.

3 — Three Things Wrong with Low Energy Prices

There are three things wrong with low energy prices. One, low energy costs leave nothing in the budget for efficient design. Really, if something is cheap, the engineer is obligated to waste it. Thus, two, low energy prices actually subsidize waste and pollution. Certainly, energy costs (and many other costs, as well) wouldn't be as low as they are currently if the full cost of cleanup were reflected in their price. But worst of all, third, low energy prices make fools of people by ruining their appreciation of the consequences of their behavior. By disconnecting behavior from its consequences, people get what is scientifically called *spoiled*. They get the oddest notions, having never had any discipline. For example, people get an astonishing sense of entitlement to waste. Even people who ought to know better...remember, consequences shape behavior. If energy has always been "too cheap to matter," considerable confusion results about what to do about consumption, waste and pollution. Then you find appalling and pathetic situations like environmentalists opposing wind power.

There is some question, as I say, whether renewable energy is a religion or a business. Market penetration is, of course, the problem with the religious model. It's just not ambitious enough. There are relatively few believers, compared to the number of potential customers, regardless of creed. Thus the business model is more likely to get systems up and running. Treating renewable energy like a business will do a lot more good, and is obviously the way of the future, should we be so fortunate as to get one.

Here's a story about energy conservation. When I was a boy, my parents bought a house in a place with a mild climate in Oregon. It came with a deal with the power company, a "Gold Medallion Home" contract for low cost energy, instead of insulation. For a modest three bedroom ranch style house in a mild climate, we used almost 4,000 kilowatt-hours a month. Since we were only paying about half a cent a kilowatt-hour, this was not a problem. Of course, today without this contract, electricity consumption in this home would be a problem.

The important thing to know is that the chapters in *Serious Microhydro* have shown just how easy it is to use renewable energy. People who are no longer the recipients of subsidies for their wasteful ways become enthusiasts of efficiency, just like their German relatives who are said to bore you with their stories of just how efficient their furnace is. There's nothing like paying the full cost for power to provide an education in energy efficiency.

Low prices, not character flaws, make waste possible. But the best way to get a swift and sometimes brutal introduction into the realities of energy politics is to develop a microhydro project. Whether your own, or by proxy through a renewable energy cooperative, or through microfinance, energy farming is without a doubt the best possible renewable energy education at all levels.

4 — Appropriate Incentives

When people have low energy costs, the proper kind of incentives for renewables may not appear obvious. Understanding how incentives work takes experience with renewable energy that just isn't available in our wasteful society, and so informed consent about energy policy is hardly possible. There are few opportunities for firsthand experience with renewable energy technology. That's why Friends of Renewable Energy BC (FOREBC) called their lecture series *Experiencing Renewable Energy*.

Our society has failed to do very much about energy and pollution since the first Earth Day in 1970. Complicated mental processes must be at play to keep us from succeeding at the relatively easy project of providing ourselves with enough clean energy. After all, we live in a world where plenty of energy falls on the roof of a house to not only power it, but drive the owner to work as well.

Let's just say that renewable energy should be considered a business much like any other. In a market study FOREBC did with solar pool heating, renewable energy was seen to act like the diffusion of an innovation in society. Microhydro is definitely such an innovation. Here's how it goes: the diffusion of an energy-saving technology into society depends upon its payback, its *return on investment*. Thus, renewables follow more or less the *payback acceptance curve.*

Basically, this means that if the payback is fairly long, like five years, there won't be much market acceptance, say 10% or so. On the other hand, as the payback improves, so does the market penetration. If the payback is a year, then the acceptance is high, like 90%. The two extremes connect with a reasonably straight line. This means that the effects of the incentive can be roughly predicted, or at least anticipated.

Some implications for and against various incentives become obvious by thinking about the payback acceptance curve. A rebate must be generous indeed to make a significant change in the payback time. By contrast, long range, low interest finance can easily offer a situation of immediate positive cash flow, or zero payback time. When the amortization time exceeds the payback time, the cost of servicing the loan is less than the cost of energy. Thus, a client is money ahead from the moment that the system begins working.

This kind of opportunism is exceedingly important when dealing with the less than affluent as well as the affluent North American. Where people make a couple of dollars a day, a significant portion of their income goes for kerosene or candles. They end up paying a lot for the poor service offered by kerosene lamps. Electricity offers a big step up in productivity and safety that may even exceed its financial rewards. It doesn't take very many house fires to pay for a modest turbine.

Other advantages of financing include the fact that regular payments constitute an incentive to keep the system running. By contrast a rebate is paid when a system is sold, and there is no such incentive. Incentives that help keep systems running have to be seen as superior. And surely financing is cheaper than a rebate program.

Although lack of financing is a significant and well-known barrier to developments of all sorts, there is nothing new about banks being unable to meet local needs. While it is generally agreed that greenhouse gas reduction is a good thing, poorer countries quite rightly balk at the expense. After all, servicing the appetites of the developed world generates much of the greenhouse gas in the world.

5 – Streamworks

Streamworks is a project to provide microhydro development with the opportunity of microfinance. Most microhydro sites remain unrecognized and undeveloped. This project can provide an excellent opportunity to provide essential services to remote areas, while producing easily documented greenhouse gas savings precisely in those areas, developing countries, where it is most difficult to do so. Financing is the missing factor in so many situations.

The proposal is to match potential microhydro clients with financing provided by more affluent enthusiasts.

To start, just do what I did, and go to a microfinance website such as kiva.org. Once there, let Bill Clinton pitch the idea of microfinance to you. The idea is pretty simple. People from the developed world loan — not give, loan — selected clients a sum of money to get something they need to better their lives. As can be seen from the work of ESMAP (Chapters 30 and 47), it is possible to deliver microhydro technology in the price range of the few hundred dollars that makes up the typical microfinance loan.

The costs of living without electricity are quite high. According to Lighting Africa, a kerosene light emits about a ton of CO_2 per year.[1] Worse, impoverished people spend 10–15% of their income on fuel for lighting. Our first microhydro project took years to get going, and so I lived many winters without electric lighting. I have experienced just how much electric lighting increases productivity. Of course, electricity does many things, but lighting is the most valuable.

It is easy enough to find significant undeveloped microhydro potential because of the lack of appropriate incentives. Even where some incentives may exist, the financing option is often absent. I sold microhydro systems for many years. Even in Canada, I could never get the kind of financing for potential clients that such a great environmental technology deserves. This situation can only be more acute where people are poor. On the other side, in our work with the nonprofit sector, we have found that there is a significant desire to get involved with renewable energy that just isn't being met in our society. For one thing, unlike solar, not everyone has access to a microhydro site. When Friends of Renewable Energy BC gave its lecture series, *Experiencing Renewable Energy*, the lecture on "Renewable Energy Co-operatives" was one of the most enthusiastically attended. We learned from this lecture that ownership of a system is not the only way to get involved with renewables. Renewable energy cooperatives have been successful because they offer a way to become involved and gain experience, without the full responsibility and cost of being an owner/operator.

Streamworks would like to offer a micro-finance structure for microhydro projects in Canada and elsewhere. FOREBC has good connections all over the world in the micro-hydro business that could steer potential clients our way, and help them qualify. And, as someone who loaned money to the program, you would get your money back in the end. For further information, contact FOREBC at forebc@shaw.ca.

Endnotes

Introduction

1. There are finally some good introductory texts generally available. For example, there's my first book, *Microhydro: Clean Power From Water* (New Society, 2003) or Natural Resources Canada's *Micro-Hydro Systems: A Buyer's Guide*. [online]. [cited April 26, 2010]. canmetenergy-canmetenergie.nrcan-rncan .gc.ca/eng/renewables/publications/micro hydro_systems.html. Another source used in this collection is Natural Resources Canada's *RETScreen Clean Energy Project Analysis Software* ([online]. [cited December 10, 2009]. retscreen.net). The discussion forum at Microhydropower.net ([online]. [cited December 10, 2009]. microhypropower.net) will also offer lots of introductory information and gossip for the beginner. In this book, many of the case studies offer their sometimes contradictory views of what introductory material is necessary.

Chapter 1: Sustainable Skiing

1. Vijay V. Vaitheeswaran. *Power to the People: How the Coming Energy Revolution Will Transform an Industry, Change Our Lives, and Maybe Even Save the Planet*. Farrar Straus and Giroux, 2003.
2. For further information:
   Auden Schendler, Director of Environmental Affairs, Aspen Skiing Company,
   P.O. Box 1248, Aspen, CO 81612,
   (t) 970-300-7152,
   (f) 970-300-7154,
   (e) aschendler@aspensnowmass.com,
   (w) aspensnowmass.com/environment (author).
   Brett Bauer, Canyon Hydro Inc.,
   P.O. Box 36, Deming, WA 98224,
   (t) 360-592-2235,
   (f) 360-592-2235,
   (e) turbines@canyonhydro.com,
   (w) canyonhydro.com (Pelton turbine and generator).
   Dan Batdorf, Bat Electric,
   20200 Charianne Drive,
   Redding, CA 96002,
   (t) 530-221-1336,
   (f) 530-221-3496,
   (e) batelecinc@aol.com (controls and switchgear).
   Pat Costello, Costello and Co.,
   405 Park Avenue, Suite E-6,
   Basalt, CO 81621,
   (t) 970-927-1421,
   (f) 970-927-2008 (contractor for powerhouse).
   Robert Gardner, Holy Cross Energy,
   P.O. Drawer 2150,
   Glenwood Springs, CO 81602,
   (t) 970-945-5491,
   (f) 970-945-4081,
   (e) bgardner@holycross.com,
   (w) holycross.com (radio-link remote terminal unit which provides generator output and bill info).
   Brian Mitchem, Mountain Peak Controls,
   P.O. Box 1550, Paonia, CO 81428,
   (t) 970-527-2444,
   (e) bmitchem@mpcontrols.com,

(w) mpcontrols.com (system automation controls).

Mike Hoffman, TPE–Twin Peaks Electric,
  145 Cheyenne Avenue,
  Carbondale, CO 81623,
  (t) 970-963-1021,
  (f) 970-963-0958,
  (e) twinpekselec@aol.com (electrician).

Mark Gressett, Gressett Excavation,
  510 Sopris Creek Road., Basalt, CO 81621,
  (t) 970-948-4686 (excavation).

Charles Brugger, Advanced Mechanical Services,
  P.O. Box 33237, Denver, CO 80233,
  (t) 303-818-5434,
  (e) advmech1@aol.com (laser alignment and turbine installation).

Tom Golec, Ruedi Creek Water and Power LLC,
  15401 Fryingpan Road, Basalt, CO 81621,
  (t) 970-927-4212,
  (e) golec@msn.com (project consultant).

Randy Udall, Community Office for Resource Efficiency (CORE),
  P.O. Box 9707, Aspen, CO 81612,
  (t) 970-544-9808,
  (e) rudall@aol.com,
  (w) aspencore.org (project consultant).

Chapter 2: From Water to Wire

1. For further information:
Peter Talbot,
  18875 124 A Avenue,
  Pitt Meadows, BC, V3Y 2G9 Canada,
  (t) 604-465-0927,
  (e) ptalbot@rptelectronics.com,
  (w) microhydro.ca, homepower.ca (author).

Malibu Club,
  P.O. Box 49,
  Egmont, BC, V0N 1N0 Canada,
  (t) 604-883-2582,
  (f) 604-883-2082,

(e) info@malibuclub.com,
  (w) malibuclub.com.

Dependable Turbines Ltd.,
  17930 Roan Place,
  Surrey, BC, Canada, V3S 5K1
  (t) 604-576-3175,
  (f) 604-576-3183,
  (e) sales@dtlhydro.com (turbine manufacturer).

Thomson and Howe,
  Site 17, Box 2, S.S. 1,
  Kimberley, BC, V1A 2Y3 Canada,
  (t) 250-427-4326,
  (f) 250-427-3577,
  (e) thes@cyberlink.bc.ca,
  (w) smallhydropower.com/thes.html (small hydro controls).

KWH Pipe (Canada) Ltd.,
  Unit 503B, 17665 66A Avenue,
  Surrey, BC, V3S 2A7 Canada,
  (t) 800-668-1892 or 604-574-7473,
  (f) 604-574-7073,
  (e) sales@kwhpipe.ca,
  (w) kwhpipe.ca (HDPE pipe).

Chapter 3: Hydro Power

1. For further information:
Paul P. Craig, College of Engineering,
  University of California, Davis 95616,
  (t) 916-752-1782 (author).

Robert Mathews has over 25 years of hands-on experience, including teaching, in renewable energy in both developed and developing countries, specializing in microhydro and photovoltaics. In British Columbia, he was owner/manager of Energy Alternatives and Appropriate Energy Systems and is still part owner of a one-megawatt hydro independent power producer in Southeastern BC. Since 2003, he has been based in Nicaragua, Central America, working on renewable energy and potable water supply projects for rural communities. Contact Bob in Managua at (e) rmathews@ibw.com.ni or (t) 011-505-278-6552 (author).

Chapter 4: Powerful Dreams

1. For further information:

Juliette and Lucien Gunderman, Crown Hill Farm,
18155 SW Baker Creek Road,
McMinnville, OR 97128,
(t/f) 503-472-5496,
(e) crownhillfarm@onlinemac.com (authors).

Canyon Industries, Inc., Brett Bauer,
5500 Blue Heron Lane,
Deming, WA 98224,
(t) 360-592-5552,
(f) 360-592-2235,
(e) CITurbine@aol.com,
(w) canyonindustriesinc.com (turbines).

Bat Electric, Inc., Dan Batdorf,
20200 Charlanne Drive,
Redding, CA 96002,
(t) 530-221-1336,
(f) 530-221-3496,
(e) BATELECINC@aol.com (co-generation control panel).

Inertia Controls, Inc., Darin Malcolm,
381 S. Redwood, Canby, OR 97013,
(t) 503-266-2094,
(f) 503-266-1152,
(e) darinm@canby.com,
(w) saftronics.com/pages/INRT.htm (temperature sensor and valve actuator panel).

Familian Northwest,
2979 N. Pacific Highway,
Woodburn, OR 97071,
(t) 866-537-7635 or 503-982-6141,
(f) 503-982-1106,
(w) familiannw.com (pipeline, valves, plumbing).

Precision Controls, Yvonne Vonderaye,
7110 SW 33rd Avenue,
Portland, OR 97219,
(t) 800-441-8246 or 503-245-7062,
(f) 503-245-4825,
(e) yvonnev@teleport.com,
(w) ifmefector.com (temperature sensors and controls).

Farnham Electric, Dennis McGill,
1050 NE Lafayette Avenue,
McMinnville, OR 97128,
(t) 503-472-2186,
(f) 503-472-4042,
(e) roberto@farnhamelectric.com (electrical contractor).

Chapter 5: Small AC

1. Natural Resources Canada. *RETScreen Clean Energy Project Analysis Software.* [online]. [cited December 28, 2009]. retscreen.net/ang/clean_energy_project_analysis.php.

2. Scott L. Davis. *Microhydro: Clean Power from Water.* New Society, 2003;

Steve Maxwell. "Homestead Hydropower: Harness the power of flowing water for clean, sustainable home electricity." *Mother Earth News* #208 (February/March 2005). [online]. [cited December 28, 2009]. motherearthnews.com/Renewable-Energy/2005-02-01/Hydroelectric-Power-Systems.aspx.

Chapter 6: Kennedy Creek

1. General contact info: 2033 Ti Bar Road, Somes Bar, CA 95568, (t) 916-469-3349 (owners live along this road).

2. For further information:

Richard Perez,
P.O.B. 130, Hornbrook, CA 96044,
(t) 916-475-3179 (author).

Harris Hydroelectric,
632 Swanton Road,
Davenport, CA 95017,
(t) 408-425-7652 (DC hydroelectric turbines).

Chapter 7: Independent Power

1. For further information:
   David Palumbo operates as Independent
   Power & Light,
   RR#1, Box 3054, Hyde Park, VT 05655,
   (t) 802-888-7194.

Chapter 8: Hydro Oz

1. See Chapter 34 of this book for a review of *The Death of Ben Linder*.
2. For further information:
   Malcolm Terence,
   (t) 831-420-1373,
   (e) mterence@sccs.santacruz.k12.ca.us
   (author).
   Don Harris, Harris Hydroelectric,
   632 Swanton Road, Davenport, CA 95017,
   (t) 831-425-7652.

Chapter 9: Oregon Trading Post

1. John Schaeffer. *Real Goods Solar Living Source Book: Your Complete Guide to Renewable Energy Technologies and Sustainable Living*. Special 30th Anniversary Edition. Gaiam Real Goods, 2007.
2. For further information:
   John Bethea,
   P.O. Box 903, Heinz OR 97738,
   (t) 541-573 4428 (author).

Chapter 10: A Working Microhydro

1. For further information:
   Harry and Marlene Rakfeldt,
   1211 Colestin Road, Ashland, OR 97520-9732 (author/owners)
   Heart Interface,
   811 1st Avenue, South Kent, WA 98032,
   (t) 206-859-0640 (inverters)
   Don Harris
   632 Swanton Road, Davenport, CA 95017 (maker of Harris turbines)
   Burkhardt Turbines
   1372 A South State Street,

Ukiah, CA 95482,
(t) 707-468-5305 (supplier of Harris turbines (packaged systems))
Renewable Energy Controls,
P.O.B. 1436, Ukiah, CA 95482,
(t) 707-462-3734 (voltage regulators)
Trojan Battery Company,
12380 Clark Street,
Santa Fe Springs, CA 90670,
(t) 800-423-6569
(outside CA) 213-946-8381 (CA)
Photron, Inc.,
149 N Main Street, Willits, CA 95490,
(t) 707-459-3211 (voltage regulator)
United States Plastic Corp.,
1390 Nuebrecht Road, Lima, OH 45801,
(t) 419-228-2242 (info) (polyethylene drums (15-55 gal) comply with FDA regulations for potable water and food storage (page 110 of 1987 catalog))
and
Consolidated Plastics Co. Inc.,
1864 Enterprise Parkway, Twinsburg, OH 44087 (page 18 of 1987 catalog)
Both catalogs may be of value to anyone in need of a variety of plastic containers and connectors/hoses.
Electron Connection Ltd.,
P.O.B. 442, Medford, OR 97501,
(t) 916-475-3179. (*The Complete Battery Book*, a compilation of information about batteries and their upkeep. This firm also designs, sells and installs complete home power systems.)
Homestead Electric,
P.O.B. 451, Northport, WA 99157,
(t) 509-732-6142 (Dave Johnson owner/consultant — hydro and solar power systems, inverters, radiotelephones)

Chapter 11: Living With Lil Otto

1. For further information:
   Dr. Hugh Spencer, Cape Tribulation Tropical

Research Station,
PMB 5, Cape Tribulation,
Queensland 4873 Australia,
(t/f) 61 70 98 00 63 (author).
Dr. Hugh Spencer is the Scientific Director, and with his wife Brigitta, the founder of the Cape Tribulation Tropical Research Station, the only research facility in the Australian wet tropical lowland rainforests. When not playing with alternative energy, he works with the local giant flying foxes or fruit bats, or develops radio-tracking technology.

An earlier version of this article appeared in the Australian alternative technology magazine *Soft Technology* (#53, 1995).

Chapter 12: Waterpower in the Andes

1. For further information:
Ron Davis,
   Campo Nuevo, Casilla 4365,
   La Paz, Bolivia,
   (t/f) 591 2 350409,
   (e) cnsorata@ceibo.entelnet.bo,
   (w) zuper.net/camponuevo.

Chapter 13: Zen and the Art of Sunshine

1. For further information:
Philip Squire, Zen Mountain Center,
   P.O. Box 43,
   Mountain Center, CA 92561,
   (t) 909-659-5272,
   (f) 909-659-3275,
   (e) shinko@zmc.org,
   (w) www.zmc.org (author).
Alternative Solar Products,
   27412 Enterprise Circle W, Suite 101,
   Temecula, CA 92590,
   (t) 800-229-SOLAR or 909-308-2366,
   (f) 909-694-1458,
   (e) mark@alternativesolar.com,
   (w) www.alternativesolar.com (system components).

Harris Hydroelectric,
   632 Swanton Road, Davenport, CA 95017,
   (t) 831-425-7652 (hydroelectric turbine and consultation).

Chapter 14: Power to the People

1. Contributors to the project:
Alternative Energy Engineering,
   Box 39HP, Redway, CA 95560,
   (t) 707-923-2277.
Earth Lab,
   358 S Main Street, Willits, CA 95490,
   (t) 707-459-6272.
Harris Hydroelectric,
   632 Swanton Road, Davenport, CA 95017,
   (t) 408-425-7652.
Integral Energy Systems,
   105 Argall Way, Nevada City, CA 95959,
   (t) 916-265-8441.
and countless groups and individuals who helped in one way or another.

Chapter 15: A Batteryless Utility Intertie Microhydro System

1. My thanks to Mark McCray of RMS Electric, Dean VanVleet of Trace Engineering, Ed Hall and Chris Badger at AES, Don Harris of Harris Hydro and Derek Veenhuis and Dennis Ledbetter of APC for their help in figuring out how to handle the dump load diversion.

2. For further information:
Kurt Johnson, The Solar Guys,
   99 Hannah Branch Road,
   Burnsville, NC 28714,
   (t) 800-614-1484 or 828-675-9866,
   (f) 828-675-4555,
   (e) kurtj12@aol.com,
   (w) solarguys.com (author).
Paul Hoover,
   (t) 828-675-5393,
   (e) Avocet365@aol.com (author).
Clara "Kitty" Couch,
   Route 8 Box 915, Burnsville, NC 28714,

(t) 828-675-5608,

(e) ckcouch@ioa.com (system owner).

Danny Honeycutt, IFIWASYOU Construction,
P.O. Box 36, Bakersville, NC 28705,

(t/f) 828-675-9144 (electrician).

French Broad Electric Membership Corporation, Charles Tolley,
P.O. Box 9, Marshall, NC 28753,

(t) 800-222-6190 or 828-649-2051,

(f) 828-649-2989,

(e) charles.tolley@frenchbroad.ncemcs
.com (utility).

Aquadyne Inc.,
P.O. Box 189, Healdsburg, CA 95448,

(t) 707-433-3813 or 303-333-6071,

(f) 707-433-3712,

(e) RKWEIR@aol.com,

(w) hydroscreen.com (Aqua Shear screen).

RMS Electric Inc., Mark McCray,
1844 55th St., Boulder, CO 80301,

(t) 800-767-5909 or 303-444-5909,

(f) 303-444-1615,

(e) memc@rmse.com,

(w) rmse.com (design consultation).

Trace Engineering, Dean VanVleet,
5916 195th St. NE, Arlington, WA 98223,

(t) 360-435-8826, ext. 2220,

(f) 360-435-2229,

(e) inverters@traceengineering.com,

(w) traceengineering.com (C-40 charge
controller and consultation).

Advanced Energy Systems, Inc. (AES),
P.O. Box 262, Wilton, NH 03086,

(t) 603-654-9322,

(f) 603-654-9324,

(e) info@advancedenergy.com,

(w) advancedenergy.com (GC 1000 inverter, AM 100 inverter monitor).

Harris Hydro, Don Harris,
632 Swanton Road, Davenport, CA 95017,

(t) 831-425-7652.

Applied Power Corp., Derek Veenhuis and
Dennis Ledbetter,

P.O. Box 339,
Redway, CA 95560,

(t) 800-777-6609 or 707-923-2277,

(f) 800-777-6648 or 707-923-3009,

(e) info@appliedpower.com,

(w) solarelectric.com (dump load and
consultation).

Chapter 17: Mini Hybrid Power System

1. For further information:

Sam Vanderhoof and Ron Kenedi, Photocomm, Inc., Independent Power Division,
930 Idaho Maryland Road, Grass Valley,
CA 95949. They manufacture the Hydro-
Charger I™. We received valuable advice,
great components and excellent service
after the sale from Sam and Ron.

Jon Hill, Integral Energy Systems, 105 Argall
Way, Nevada City, CA 95959. Jon has
helped us numerous times with products
and advice. His workshops on alternative
energy and hot water production are
great!

Trace Engineering, 5917 195th NE, Arlington,
WA 98223. We own the Trace Model 1512
Inverter. Everybody knows the Trace is the
greatest.

Jim Cullen. *How To Be Your Own Power Company: Low Voltage, Direct Current, Power
Generating System*. Van Nostrand, 1980.
This guide to low voltage technology and
alternative power design was invaluable in
developing our system.

Editors of Sunset and Southern Living. *Basic
Home Wiring—Illustrated*. Lane, 1989.
Introduction to basic electric terms, system
design and construction. For the do-it-
yourselfer.

David Cooperfield "Electrical Independence"
booklet series. Well-Being Productions,
P.O. Box 757, Rough & Ready, CA 95975.
This series of booklets on electrical independence is most helpful.

Chapter 18: Hydro Power Done Dirt Cheap

1. For further information:
   Eileen Loschky nee Puttre,
   (e) eloschky@comcast.net (author).
   Stephen M. Gima has died since the article was written.
   Alternative Energy Engineering,
   P.O. Box 339, Redway, CA 95560,
   (t) 800-777-6609 (Pelton wheel).

Chapter 19: A Microhydro Learning Experience

1. For further information:
   Louis Woofenden and Rose Woofenden,
   P.O. Box 1001, Anacortes, WA 98221,
   (f) 360-293-7034,
   (e) kc7hdc@arrl.net (authors)
   Jo Hamilton, Solar Plexus,
   130 W. Front Street, Missoula, MT 59802,
   (t/f) 406-721-1130,
   (e) solplex@montana.com,
   (w) montana.com/solplex (author).
   Solar Energy International,
   P.O. Box 715, Carbondale, CO 81623,
   (t) 970-963-8855,
   (f) 970-963-8866,
   (e) sei@solarenergy.org,
   (w) solarenergy.org.
   Bob Mathews, in Managua, Nicaragua at
   (e) rmathews@ibw.com.ni.
   John Heil, Dyno Battery,
   4248 23rd Avenue W., Seattle, WA 98199,
   (t) 877-DYNO-BAT or 206-283-7450,
   (f) 206-283-7498,
   (e) dyno01@aol.com,
   (w) dynobattery.com.
   Energy Systems and Design,
   P.O. Box 4557,
   Sussex, NB, E4E 5L7 Canada,
   (t) 506-433-3151,
   (f) 506-433-6151,
   (e) hydropow@nbnet.nb.ca,
   (w) microhydropower.com.

Dan New, Canyon Industries,
   5500 Blue Heron Lane,
   Deming, WA 98244,
   (t) 360-592-5552 ,
   (f) 360-592-2235,
   (e) CITurbine@aol.com,
   (w) canyonindustriesinc.com.
Chris Soler, Soler Hydro-Electric,
   18067 Colony Road, Bow, WA 98232,
   (t) 360-724-5111.

Chapter 20: Hydro — New England Style

1. For further information:
   Bill Kelsey,
   5 Weber Road, Sharon, CT 06069,
   (t) 860-364-0288 (author).
   Don Harris, Harris Hydroelectric,
   632 Swanton Road,
   Davenport, CA 95017,
   (t) 831-425-7652.
   Larry Riley, Riley Electric,
   116 Point-of-Rocks Road,
   Falls Village, CT 06031,
   (t) 860-824-0859 (electrician).
   BP Solar,
   630 Solarex Court, Frederick, MD 21703,
   (t) 800-521-7652 or 410-981-0240,
   (f) 410-981-0278,
   (e) info@bpsolar.com,
   (w) bpsolar.com.
   New England Solar Electric Inc.,
   401 Huntington Road,
   Worthington, MA 01098,
   (t) 800-14-4131 or 413-238-5974,
   (f) 413-238-0203,
   (e) nesolar@newenglandsolar.com,
   (w) newenglandsolar.com.
   Xantrex Technology Inc.,
   5916 195th St. NE, Arlington, WA 98223,
   (t) 800-670-0707 or 360-435-8826,
   (f) 360-435-3547,
   (e) info@xantrex.com,
   (w) xantrex.com.

Chapter 21: Remote Power and Amateur Radio

1. For further information:
   Peter Talbot (VE7CVJ),
      18875 124 A Avenue,
      Pitt Meadows, BC, Canada V3Y 2G9,
      (t) 604-465-0927,
      (e) ptalbot@vcn.bc.ca (author).

Chapter 22: Been There, Done That

1. More technical details on this 12 to 24 conversion can be found in the article in *Home Power* Issue #41 that followed this one. See *Home Power* Issue #33, page 84 for a review of the Voltmaster.

2. For further information:
   Richard Perez, c/o *Home Power*,
      P.O. Box 520, Ashland, OR 97520,
      (t) 916-475-3179 (author).
   Bob-O Schultze and Kathleen Jarschke-Schultze, Electron Connection,
      P.O. Box 203, Hornbrook, CA 96044,
      (t) 916-475-3402,
      (f) 916-475-3401 (system owners).

Chapter 23: The Ten Kinzel/Kingsley Rules

1. For further information:
   Terry Kinzel and Sue Ellen Kingsley,
      R.R.1 Box 68, Hancock, MI 49930
      (authors).

Chapter 24: Ultra-Low Head Hydro

1. Books to look for:
   Carl C. Harris and Samuel O. Rice. *Power Development Of Small Streams: A Book For All Persons Seeking Greater Comfort And Higher Efficiency In Country Homes, Towns And Villages*. 1920; reprint, Kessinger, 2007.
   James Leffel and Co. *Construction of Mill Dams*. 1881; Reprint; Noyes Press, 1972.
   Rodney Hunt. *Water Wheel Catalogue #44*. (author note: the best — check out the engineering section).
   Any catalogs printed by James Leffel and Company, S. Morgan Smith Company, Fitz Water Wheel Company, Holyoke Machine Company, Dayton Globe Manufacturing Company.

2. For further information:
   Cameron MacLeod,
      P.O.B. 286, Glenmoore, PA 19343,
      (t) 215-458-8133 (author).

Chapter 25: Rolling Thunder

1. For further information:
   Richard Perez, c/o *Home Power*,
      P.O.B. 130, Hornbrook, CA 96044,
      (t) 916-475-3179 (author).
   Stuart Higgs,
      7104 Old Shasta Road, Yreka, CA 96097,
      (t) 916-842-6921 (system designer, installer and operator).

Chapter 26: Handmade Hydro Homestead

1. For further information:
   Bob-0 Schultz, Electron Connection,
      (e) bob-o@electronconnection.com (author).
   Matt and Roseanne Olson,
      Methodist Creek, Forks of the Salmon, CA 96031 (owner/operators).

Chapter 27: Choosing Microhydro

1. For further information:
   Jeffe Aronson, RMB 2143,
      Anglers Rest, Victoria 3898, Australia,
      (t) 03-5159-7252,
      (e) effe@tpg.com.au.
   In the US: c/o Echo River Trips,
      6529 Telegraph Avenue, Oakland, CA 94609,
      (t) 800-652-3246 or 510-652-1600,
      (f) 510-652-3987,
      (w) echotrips.com (author).
   Solar Australia, Melbourne,
      Unit 1, 15 Nicole Close,
      Bayswater North VIC 3153,

(t) 03-9761-5877,

(f) 03-9761-7789,

(e) sales@solaraustralia.com.au
(inverter).

Australian Energy Research Laboratories P/L
(AERL), Stuart Watkinson,

2 Roslyn Court, North Tamborine 4272,
Queensland, Australia,

(t) 07-5545-0177,

(f) 07-5545-0866,

(e) aerl@hotkey.net.au,

(w) windsun.com/ChargeControls/maxi
mize.htm (original charge controller and
dump load).

Plasmatronics,

14 Gipps Street, Collingwood, Victoria,
Australia, 3066,

(t) 03-9486-9902,

(f) 03-9486-9903,

(e) admin@plasmatronics.com.au,

(w) plasmatronics.com.au (PL-20 charge
controller and metering).

Saft Nife Power Systems Australia P/L,
Gary Piper,

18 May Road, Lalor,

Victoria, Australia 3075,

(t) 03-9465-5734,

(f) 03-9465-6213,

(e) gary.piper@saft.alcatel.com.au,

(w) saftbatteries.com (batteries).

Chapter 28: Water Rites

1. For further information:

Jeffe Aronson (jeffe@tpg.com.au) is a 33-year
veteran river guide, having rowed and
paddled in South America, Australia,
Africa and the United States. He's owned
a natural foods store, pioneered river trips
in the Grand Canyon for disabled persons
and directed the historic restoration of
downtown Flagstaff, Arizona (author).

Energy Systems and Design,

(t) 506-433-3151,

(w) microhydropower.com (turbine manufacturer).

Microhydro listserv,

(e) microhydro-subscribe@yahoogroups
.com,

(w) groups.yahoo.com/group/microhydro
(information and help with microhydro
systems).

Chapter 29: 240 VAC Direct Drive Hydro

1. For further information:

John Hermans,

320 Bellbird Road, Clifton Creek,
Victoria 3875

(e) john.g.hermans@gmail.com (author).

J. A. Brebner PL, 74 Second Street, Ashbury,
N. S. W. 2193, Australia,

(e) pfa@ozemail.com.au,

(w) pactok.net.au/docs/apace/home.htm
(Australia microhydro governor).

Chapter 31: One Dam On Its Way

1. turbin@cargo-kraft.se.

2. For further information:

C. MacLeod & Co.,

2131 Harmonyville Road,

Pottstown, PA 19465,

(t) 610-469-1858,

(f) 610-469-1859,

(e) microhydro@comcast.net (author).

Chapter 33: Small Water Power Siting

1. For further information:

Author Paul Cunningham (hydropow@
nbnet.nb.ca) is CEO of Energy Systems
and Design (microhydropower.com). He
manufactures water machines and lives on
hydro power.

Chapter 34: Review of The Death of Ben Linder

1. Joan Kruckewitt. *The Death of Ben Linder:
The Story of a North American in Sandinista*

Nicaragua. Seven Stories Press, 2001. 20% discount for website purchases — sevenstories .com.

2. For further information:
Chris Greacen and Arne Jacobson,
  310 Barrows Hall ERG, UC Berkeley,
  CA 94720-3050,
  (e) cgreacen@socrates.berkeley.edu,
  (e) arne@socrates.berkeley.edu (authors).

Chapter 35: Microhydro Power in the Nineties

1. For further information:
Paul Cunningham, Energy Systems and
  Design,
  P.O.B. 1557,
  Sussex, New Brunswick, Canada E0E 1P0,
  (t) 506433-3151,
  (e) hydropow@nbnet.nb.ca,
  (w) microhydropower.com (author).
Barbara Atkinson,
  1401 Acton Crescent,
  Berkeley, CA 94702 USA,
  (t) 510-841-2373 (author)
Robert Mathews, Appropriate Energy Systems,
  (e) rmathews@ibw.com.ni (reviewer).

Chapter 36: Soft Starting Electrical Motors

1. For further information:
Jim Forgette, Wattevr Works,
  P.O.B. 207,
  San Andreas, CA 95249,
  (t) 209-754-3627 (author).

Chapter 37: Microhydro Intake Design

1. For further information:
Jerry Ostermeier owns Alternative Power and
  Machine in Grants Pass, Oregon,
  (t) 541-476-8916,
  (e) altpower@grantspass.com,
  (w) apmhydro.com. He has been designing and installing microhydro and off-grid
  power systems since 1979. He also manufactures a user-friendly, residential-scale
  microhydro turbine (author).

Amiad Filtration Systems,
  (w) amiadusa.com (in-line screens).
Hydroscreen Co. LLC,
  (w) hydroscreen.com (Coanda-effect
  screens).

Chapter 39: The Hydro's Back

1. For further information:
Kathleen Jarschke-Schultze
  (c/o *Home Power* magazine,
  P.O. Box 520,
  Ashland, OR 97520,
  (e) kathleen.jarschke-schultze@home
  power.com)
  is starting to dabble in viticulture at
  her home in Northernmost California
  (author).
Energy Systems and Design, Ltd.,
  P.O. Box 4557,
  Sussex, NB, Canada E4E 5L7,
  (t) 506-433-3151,
  (f) 506-433-6151,
  (e) hydropow@nbnet.nb.ca,
  (w) microhydropower.com (Stream Engine
  microhydro generators).

Chapter 40: The Village or the House

1. For further information:
Chris Greacen, Energy and Resources Group,
  310 Barrows Hall #3050,
  University of California,
  Berkeley, CA 94720-3050,
  (t) 510-643-1928,
  (f) 510-642-1085,
  (e) cgreacen@socrates.berkeley.edu
  (author).
Jeevan Goff, Lotus Energy,
  PO Box 9219,
  Bhatbhateni Dhunge Dhara,
  Kathmandu, Nepal,
  (t) +977 (1) 418 203,
  (f) +977 (1) 412 924.
  (e) jeevan@lotusnrg.com.np,
  (w) lotusenergy.com (author).

Chapter 41: Twenty Years of People Power

1. For further information:
   Joe Schwartz, Home Power,
   P.O. Box 520, Ashland, OR 97520,
   (e) joe.schwartz@homepower.com,
   (w) homepower.com (author).
   Ian Woofenden,
   P.O. Box 1001, Anacortes, WA 98221,
   (e) ian.woofenden@homepower.com,
   (w) homepower.com (author).

Chapter 43: Practical Solar

1. For further information:
   Philippe Habib,
   526 View Street,
   Mountain View, CA 94041,
   (t) 650-968-8654,
   (e) habib@well.com (author).
   Xantrex Technology Inc., Distributed Residential and Commercial Markets,
   5916 195th St. NE,
   Arlington, WA 98223,
   (t) 360-435-8826,
   (f) 360-435-2229,
   (e) inverters@traceengineering.com,
   (w) tracegridtie.com/products/gridtie/suntie.html.
   Solatron Technologies Inc.,
   aka Solar On Sale,
   19059 Valley Boulevard, Suite 219,
   Bloomington, CA 92316,
   (t) 877-744-3325 or 909-877-8981,
   (f) 909-877-8982,
   (e) sales@solaronsale.com,,
   (w) solaronsale.com.
   Renewable Energy Program, California
   Energy Commission (CEC),
   1516 Ninth Street, MS-45,
   Sacramento, CA 95814,
   (t) 800-555-7794 or 916-654-4058,
   (e) renewable@energy.state.ca.us,
   (w) energy.ca.gov/greengrid.
   Pacific Gas and Electric (PG&E),
   (w) pge.com

Efrain Ornelas: Program manager for clean
air transportation,
(t) 415-972-5617
Harold Hirsh: Renewable energy department,
(t) 415-973-1305
Jerry Hutchinson: Takes care of getting your
system legally grid tied,
(t) 408-282-7345
Dave Turner: PG&E E-NET Processing
Department,
(t) 415-973-4525
Phil Quadrini: Knows tariffs forwards and
backwards,
(t) 415-973-4213.
American Water Heater Group,
P.O. Box 4056, Johnson City, TN 37602,
(t) 800-999-9515,
(w) americanwaterheater.com (Polaris
water heater).
Temp-Cast Enviroheat,
P.O. Box 94059, 3324 Yonge Street,
Toronto, ON M4N 3R1 Canada,
(t) 800-561-8594 or 416-322-6084,
(f) 416-486-3624,
(e) staywarm@tempcast.com,
(w) tempcast.com (masonry heater).
General Electric,
(t) 800-432-2572 or 770-999-7131,
(w) geindustrial.com/industrialsystems/
meter/catalog/kvmeter (TOU meter,
model KV).

Chapter 44: Beyond Net Metering

1. For further information:
   Don Loweburg, IPP,
   P.O. Box 231, North Fork, CA 93643,
   (t) 209-977-7080,
   (e) i2p@aol.com (author).
   i2p.org (IPP webpage).
   David Parker,
   (e) esco@efn.org (OSEIA President).
   Ken Bossong,
   (e) kbossong@cais.com (*Sustainable Energy Coalition Update* newsletter).

Chapter 47: Stimulating the Picohydropower Market

1. ESMAP, c/o Energy and Water Department,
   The World Bank Group,
   1818 H Street N.W.,
   Washington, D.C. 20433, U.S.A.

Chapter 49: Market Solutions

1. Paul Tolme. "How Cheap LEDs Could Efficiently Power Africa and Beyond" *Popular Mechanics* [online]. [cited April 19, 2010]. popularmechanics.com/science/earth/4248193.html?page=2.

Index

Page numbers in **bold** indicate photographs and illustrations.

About the Author

Scott L. Davis dropped out of graduate school in 1976 to work on, among other things, a village scale hydroelectric project in a remote area west of Lillooet BC. In the years that followed, he used, designed, constructed, installed, repaired and generally fooled around with the many microhydro systems that sprang up in the area.

In 1995, he and his partner, Bonnie Mae Newsmall, founded Yalakom Appropriate Technology, an award winning renewable energy project development company. Their projects were featured as case studies by the Canadian Renewable Energy Guide, in RETSCREEN — Canada's renewable energy software and in *Microhydropower Systems: A Buyers' Guide* and the *Commercial Solar Water Heating Buyers' Guide*, published by Natural Resources Canada.

It soon became clear that the significant barriers to renewable energy investment in BC and Canada could best be surmounted by a nonprofit organization. To that end, in 2000, Scotty and Bonnie Mae founded Friends of Renewable Energy BC, which continues to look for market solutions for our environmental problems. FOREBC's projects have included a self-directed tour of existing renew-

able energy technology, a renewable energy lecture series and a very successful microhydro workshop. More projects are planned, including the Streamworks project described in Chapter 49.

Today, Scotty enjoys working with Friends of Renewable Energy BC and life, especially the gardening, in Victoria BC.

If you have enjoyed *Serious Microhydro*,
you might also enjoy other

BOOKS TO BUILD A NEW SOCIETY

Our books provide positive solutions for people who
want to make a difference. We specialize in:

Sustainable Living ◆ Ecological Design and Planning

Natural Building & Appropriate Technology ◆ New Forestry

Environment and Justice ◆ Conscientious Commerce

Progressive Leadership ◆ Resistance and Community

Nonviolence ◆ Educational and Parenting Resources

New Society Publishers

ENVIRONMENTAL BENEFITS STATEMENT

New Society Publishers has chosen to produce this book on recycled
paper made with 100% post consumer waste, processed chlorine free,
and old growth free.

For every 5,000 books printed, New Society saves the following resources:[1]

45	Trees
4,087	Pounds of Solid Waste
4,496	Gallons of Water
5,865	Kilowatt Hours of Electricity
7,429	Pounds of Greenhouse Gases
32	Pounds of HAPs, VOCs, and AOX Combined
11	Cubic Yards of Landfill Space

[1]Environmental benefits are calculated based on research done by the Environmental
Defense Fund and other members of the Paper Task Force who study the environmental
impacts of the paper industry.

For a full list of NSP's titles, please call 1-800-567-6772 or check out our web site at:

www.newsociety.com

NEW SOCIETY PUBLISHERS